Clinical Applications of the *Limulus* Amoebocyte Lysate Test

Editor

Richard B. Prior, Ph.D.
Associate Professor of Internal Medicine
Division of Infectious Diseases
Department of Internal Medicine
The Ohio State University College of Medicine, and
Associate Professor of Microbiology
The Ohio State University College of Biological Sciences
Columbus, Ohio

CRC Press
Boca Raton Ann Arbor Boston

Library of Congress Cataloging-in-Publication Data

Clinical applications of the *Limulus* amoebocyte lysate test / edited
by Richard B. Prior.
p. cm.
Includes bibliographical references.
ISBN 0-8493-6209-1
1. Gram-negative bacterial infections—Diagnosis. 2. *Limulus*
test. 3. Endotoxins—Analysis. I. Prior, Richard B., 1941- .
[DNLM: 1. Endotoxins—analysis. 2. *Limulus* Test. QW 630 C6413]
RC116.G677C55 1990
616.9′20756—dc20
DNLM/DLC
for Library of Congress 90-1548
CIP

This book represents information obtained from authentic and highly regarded sources. Reprinted material is quoted with permission, and sources are indicated. A wide variety of references are listed. Every reasonable effort has been made to give reliable data and information, but the author and the publisher cannot assume responsibility for the validity of all materials or for the consequences of their use.

Direct all inquiries to CRC Press, Inc., 2000 Corporate Blvd., N.W., Boca Raton, Florida, 33431.

International Standard Book Number 0-8493-6209-1

Library of Congress Card Number 90-1548
Printed in the United States

PREFACE

Endotoxins (lipopolysaccharides [LPS]) are part of the cell wall of Gram-negative bacteria and are widely distributed in nature, including the water supplies. Yet endotoxins are usually not found in human body fluids (blood, lymph, CSF, urine, semen, ocular fluid, etc.) under conditions of good health. For Gram-negative bacteria, endotoxins are important elements in their pathogenicity. However, the exact role these substances play in disease processes, especially chronic diseases, remains unclear. Nevertheless, detecting the presence of Gram-negative bacterial endotoxins in certain clinical situations can be extremely useful to the clinician in the management of patients.

In 1964, Levin and Bang reported on the role of bacterial endotoxins on the extracellular coagulation of blood from the horseshoe crab, *Limulus polyphemus*. This reaction appeared to be an ancient defense mechanism permitting the horseshoe crab to wall off invading bacteria, usually Vibrio, found in its natural habitat, the sea. Subsequent work revealed that the substances involved in this mechanism were associated with the amoebocyte, the only circulating cellular element of the horseshoe crab. Lysate made from the amoebocytes was found to produce a gel or clot when exposed to small quantities of bacterial endotoxins. The reaction was enzymatic in nature and appeared to be triggered only by endotoxins. Thus began the *Limulus* amoebocyte lysate (LAL) test for the detection of Gram-negative bacterial endotoxins.

Since then many investigators have evaluated the LAL test in numerous clinical and nonclinical situations where the detection of endotoxins would be helpful. The first and most successful use of the LAL test was in the pharmaceutical industry. The detection of fever-producing substances — pyrogens — in radiolabeled drugs, pharmaceutical preparations, and invasive medical devices using the LAL test proved to be a very sensitive, specific, rapid, and economical method to evaluate these materials for human use. Today, numerous LAL tests are performed daily in the pharmaceutical industry in place of the costly and time-consuming rabbit fever test.

The parallel development of the LAL test in clinical medicine has not been as successful as in the pharmaceutical industry. Early reports showed conflicting results when testing blood for endotoxins and it seemed that each report, regardless of the clinical situation evaluated, used a slightly different method to perform the LAL test. In addition, there were reported variations within the *Limulus* lysate preparations. It became apparent that certain standards were needed both in the LAL methodology and in the preparation of the lysate. Today, high-quality lysate is commercially available and labeled potencies are expressed in Endotoxin Units (EU). Lysate preparations are very stable and currently can detect picogram (10^{-12} grams) quantities of endotoxin. The U.S. Food and Drug Administration (FDA) now regulates the LAL test in clinical medicine and the lysate prepared by various manufacturers.

It is the primary intention of *Clinical Applications of the Limulus Amoebocyte Lysate Test* to present the clinical application of the LAL test as reported in the literature and coalesce this information into a clinical syndrome approach. Leading authorities have contributed chapters that present (1) an introduction reviewing the clinical syndrome, its causes, epidemiology, diagnoses, etc. and states the problem of evaluating the syndrome and the need for such evaluation; (2) a brief background describing the methods currently used to evaluate the syndrome with a discussion of their pros and cons; (3) a review of the past and present status of the LAL test in evaluating the clinical syndrome including discussions about clinical utility, sensitivity, specificity, and predictive values; (4) a discussion of the future trends of the role of the LAL test in the clinical situation; and (5) a complete and current listing of the relevant literature. In addition to the clinical syndromes discussed, there are chapters that (1) present an overview of bacterial endotoxins and their clinical significance, (2) describe the horseshoe crab to familiarize the reader with this unique creature, (3) describe the various methodologies used in the LAL test, (4) describe the role of the FDA in the regulation of the LAL test, (5) describe the various allied medical applications of the LAL test, including its use in hemodialysis, (6) present an overview of the veterinary applications of the LAL test, (7) present an overview of the pharmaceutical applications of the LAL test, and (8) summarize the material presented and discuss future trends of the LAL test in clinical medicine.

Clinical Applications of the Limulus Amoebocyte Lysate Test is directed toward physicians, medical students, nurses, pathologists, toxicologists, microbiologists, public health officials, laboratory technicians, and other paraprofessionals involved in the teaching, evaluation, management, and treatment of clinical situations involving Gram-negative bacteria. It is hoped that those currently involved in using the LAL test for research purposes will find new insights and ideas in the material presented here. Likewise, itis hoped that new researchers will explore this test and perhaps report a unique finding that will be helpful to our understanding and use of this test in clinical medicine.

I wish to thank Vincent A. Spagna for his enthusiasm and great ideas as we pioneered our work in gonorrhea using the LAL test over 10 years ago. It was a very productive period for the both of us and it shows the meaning of the word "synergy" when two researchers of dissimilar backgrounds collaborate. I also wish to thank the contributors for their participation and cooperation during this project.

Richard B. Prior, Ph.D.

FOREWORD

The *Limulus* amoebocyte lysate (LAL) test for bacterial endotoxins was developed as the result of investigations of the effects of Gram-negative bacteria on the blood coagulation mechanism of the horseshoe crab, *Limulus polyphemus*. Initial studies of blood coagulation of the horseshoe crab were hampered because samples of blood, which were initially liquid, became solidly clotted by the next morning, despite the presence of various anticoagulants. I considered the possibility of bacterial contamination of the blood specimens, and when samples of blood were collected into sterile and pyrogen-free containers (the latter being attempted only because of my other studies with endotoxins), the blood remained liquid. It was then possible to demonstrate that the coagulation system, or at least some of its components, leaked from amoebocytes, the only type of circulating blood in *L. polyphemus*, and that plasma which contained cellular components could be gelled by bacterial endotoxins. Our first paper that specifically described lysates prepared from washed amoebocytes demonstrated a relationship between the concentration of endotoxin and the rate of gelation or the rate of increase of turbidity, and was published in 1968. Biological quantification of endotoxin, *in vitro*, was made possible by the recognition of the sensitivity of LAL to a wide variety of bacterial endotoxins.

This volume is the sixth book to be based on the LAL test for bacterial endotoxins. The previous five volumes were derived from international meetings concerned with *L. polyphemus* and the LAL test. A review of the five preceding volumes reveals the development of an understanding of the LAL test and its performance, and the increasing range of its utilization. In *Biomedical Applications of the Horseshoe Crab* (1979) it was incorrectly suggested that there was only a single pertinent enzyme involved in the coagulation mechanism present in amoebocyte lysate. Only one paper reported the use of chromogenic substrates in conjunction with LAL. The second book, *Endotoxins and Their Detection with the Limulus Amebocyte Lysate Test,* was published in 1982. Much attention was devoted to standardization of the LAL test and establishment of a reference endotoxin. Although there was continued discussion dealing with measurement of endotoxin with the rabbit pyrogen test vs. the LAL test, there clearly was a trend toward acceptance of this test for many applications. The third volume was entitled *Bacterial Endotoxins. Structure, Biomedical Significance, and Detection with the Limulus Amebocyte Lysate Test* (1985). Pertinently, there were only two papers which included a comparison of the LAL test with the rabbit pyrogen test. For the first time, a large section was devoted to clinical applications of the LAL test. Five papers reported the use of chromogenic substrates for its performance. The most recent of these volumes was published in 1988 and is entitled *Bacterial Endotoxins. Pathophysiological Effects, Clinical Significance, and Pharmacological Control.* The contents of this volume clearly documented the acceptance of the LAL test as the standard assay for the detection of bacterial endotoxins in a wide variety of applications, and revealed increasing interest in its use for the evaluation of endotoxemia.

An independent measure of the acceptance of the LAL test has een documented in medical libraries. In 1977, nine years after our initial paper which described the use of LAL for the detection of bacterial endotoxins, the term "*Limulus* Test" first appeared in the Cumulative Index Medicus. In 1983, the Medical Subject Headings, published by the National Library of Medicine, provided a separate line for the term "*Limulus* Test". Dr. Hochstein's chapter reviews the careful steps that led to the licensing of the LAL test by the FDA.

Overall, the LAL test has shown generally excellent correlations with many other bioassays for endotoxin, such as pyrogenicity, mitogenicity, tissue factor generation, complement activation, and chick embryo lethality. Since many of the effects of endotoxin that are measured in other bioassays are indirect, whereas the gelation of LAL is a relatively direct effect, these correlations are rather remarkable. Recent studies, based on the availability of synthetic Lipid A, seem to have clearly demonstrated the overall specificity of the LAL test for bacterial endotoxin and the requirement for certain chemical structures in the Lipid A molecule.

However, as pointed out by many of the contributors to this volume, it is important to note that the physical state of endotoxin probably plays an important role in its potential range of biological effects. With the establishment of the LAL test as the standard assay for endotoxin, we should probably no longer equate the presence of endotoxin with pyrogenicity. The term "pyrogen-free" was the result of the previous standard rabbit assay for endotoxin. Since the rabbit pyrogen assay has been essentially supplanted by the LAL test, it would seem no longer appropriate to define endotoxin by an activity that is no longer being routinely evaluated. Furthermore, it is noteworthy that pyrogenicity is probably one of the safer and better tolerated effects of bacterial endotoxins, and perhaps it is more clinically relevant to concern ourselves with the presumptive effects of endotoxin on blood coagulation, blood pressure, or complement, when evaluating the clinical significance of a positive LAL test.

There are now multiple, well-established applications for the LAL test. Pharmaceutical preparations, which must be endotoxin-free, are now routinely evaluated with this test. Such materials include both large and small volume parenterals, certain drugs, vaccines, and biological products. This test can be effectively used to determine the presence of endotoxin in short-lived radioiosotopes that are used for diagnosis. It provides an efficient method for the field testing of water supplies for contamination by coliform organisms, and, as suggested in the interesting chapter by Richter and Mikolajcik, for detection of Gram-negative bacteria and their endotoxins in milk. The feasibility of the LAL test has made it possible for the first time for research laboratories to rule out the presence of endotoxin in biologically active reagents, such as mitogens, erythropoietin and cytokines, in which the presence of contaminating bacterial endotoxin could confound results and generate misleading data. It also has been used effectively for the detection of endotoxin in plasma protein fractions, especially albumin, and for various medical devices.

The use of the LAL test for the detection of endotoxin in body fluids is just beginning to develop. Extensive data already document its usefulness for the diagnosis of Gram-negative meningitis. The thorough and critical review of this area by Nachum documents the suitability of the LAL test to diagnose or rule out the presence of Gram-negative bacterial meningitis. Seventeen of the nineteen studies identified by Nachum clearly established the very high sensitivity (543 positive LAL tests in 557 samples of CSF from which Gram-negative organisms were cultured, 97.5%) and specificity (only 1 positive LAL test in 166 specimens in which Gram-positive organisms were present, 99.4%) of this test. In contrast, the two other studies surveyed demonstrated a sensitivity of only 72% and a specificity of 87%. One of the latter two studies accounted for 6 of the 7 false-positive LAL tests reported in a total of 210 cases with Gram-positive meningitis. The importance of proper performance of the LAL test, which requires attention to detail and continuous effort to prevent contamination of reagents and equipment by endotoxin, cannot be overemphasized. Chapters by Nachum, Prior, and Alfonso and Miller indicate that the detection of endotoxin in other body fluids, i.e., urine, urethral secretions, and corneal tissue, can be of considerable diagnostic value. Interestingly, for both urine and urethral secretions, dilution of the body fluid has been successfully used to discriminate between significant and clinically unimportant evidence of endotoxin. Considering the quality and extent of the data in these clinical areas, it is surprising that the LAL test has not been utilized more widely, especially since its ability to indicate the presence of Gram-negative bacteria is not limited by prior or concurrent administration of antibiotics.

Detection of endotoxin in blood remains the greatest challenge for a test for bacterial endotoxins because of the presence of inhibitors of endotoxin in blood. Multiple techniques have been described to remove these inhibitors prior to utilization of the LAL test. We have recently compared the effectiveness of chloroform extraction (which we originally described), acid precipitation, and dilution/heating. Our comparison has demonstrated that the combination of dilution and heating of plasma is the most effective among these three techniques for the removal of inhibitors of endotoxin from blood. Unexplained are the wide ranges of dilution and heating that have been reported to be effective by different investigators. This important preparatory step requires resolution if the LAL test is to be used to its maximal advantage for the diagnosis of patients with Gram-negative sepsis.

It is important to extend utilization of the LAL test for the study of endotoxemia, now that there is belated acceptance of the concept that endotoxemia can be present without bacteremia. The conclusion by some that without a concurrently positive blood culture, a positive LAL test represents a false-positive result was never justified. Bacteremia and endotoxemia are not necessarily related phenomena, and the average concentrations of bacteria in the blood, during usually short-lived bacteremic episodes, would almost never account for the levels of endotoxin detected by the LAL test. Only by serial studies of the presence of endotoxin in the blood of patients with appropriate disorders will we be able to expand our understanding of the pathophysiologic significance of endotoxemia and correlate endotoxemia with the potentially enormous range of effects that can be produced by bacterial endotoxins. Serial studies of the presence of circulating endotoxin in appropriate disease states also should provide insights into the probable role of endotoxin normally present in the gastrointestinal tract in producing clinical syndromes characterized by "sepsis".

Acquisition of this knowledge has become more important with the advent of a new group of agents designed for the treatment of endotoxemia rather than bacteremia, i.e., anti-endotoxin antibodies. Other therapeutic modalities are likely to become available in the future and it is important that we fully understand the implications of endotoxemia.

It is important to note that the calculation of the concentration of an unknown bacterial endotoxin in a sample, using this or any other test, is critically dependent upon the standard endotoxin used and the sensitivity of the LAL. For example, the purer or more biologically active the endotoxin standard, the greater the concentration of the unknown will appear to be. Equal concentrations (by weight) of different bacterial endotoxins do not necessarily have the same biological activity in this or any other assay for endotoxin. The increasing use of chromogenic substrates in conjunction with the LAL test makes this consideration even more important, since it has been shown that the different chromogenic substrates have different rates of reaction with the enzyme or enzymes in LAL. In addition, it has been clearly demonstrated that different sources of bacterial endotoxin, at equivalent concentrations by weight, produce different rates of cleavage of a single chromogenic substrate. In my opinion, there has been inadequate consideration of these important and related aspects of the performance of the LAL test. Therefore, levels of bacterial endotoxin measured in body fluids should be considered as relative, and not interpreted as absolute values.

The LAL test is currently the most sensitive test available for the detection of bacterial endotoxin and correlates with a wide variety of other assays. This book, the first which focuses exclusively on the clinical applications of the LAL test, summarizes and evaluates an enormous body of information. Importantly, it thereby has documented and clarified the established applications of the LAL test and suggests new areas in which this test can be carefully evaluated and then productively utilized for the diagnosis of Gram-negative infection and monitoring of the effectiveness of therapy.

Jack Levin, M.D.
San Francisco

SELECTED BIBLIOGRAPHY

Cohen, E., Bang, F. B., and Levin, J., et al., Eds., *Biomedical Applications of the Horseshoe Crab* (Limulidae), Alan R. Liss, New York, 1979.

Watson, S. W., Levin, J., and Novitsky, T. J., Eds., *Endotoxins and Their Detection with the Limulus Amebocyte Lysate Test,* Alan R. Liss, New York, 1982.

Ten Cate, J. W., Buller, H. R., Sturk, A., and Levin, J., Eds., *Bacterial Endotoxins. Structure, Biomedical Significance, and Detection with the Limulus Amebocyte Lysate Test,* Alan R. Liss, New York, 1985.

Watson, S. W., Levin, J., and Novitsky, T. J., *Detection of Bacterial Endotoxins with the Limulus Amebocyte Lysate Test,* Alan R. Liss, New York, 1987.

Levin, J., Buller, H. R., Ten Cate, J. W., van Deventer, S. J. H., and Sturk, A., Eds., *Bacterial Endotoxins. Pathophysiological Effects, Clinical Significance, and Pharmacological Control,* Alan R. Liss, New York, 1988.

EDITOR

Richard B. Prior, Ph.D., is Associate Professor of Internal Medicine in the Division of Infectious Diseases, Department of Internal Medicine at The Ohio State University College of Medicine, Columbus, Ohio. He also holds a joint appointment as Associate Professor of Microbiology in the Department of Microbiology at The Ohio State University College of Biological Sciences. He received his Ph.D. from The Ohio State University in 1972 and joined the faculty in the Division of Infectious Diseases at that time.

Dr. Prior is a member of the American Society for Microbiology, the American Association for the Advancement of Science, the American Venereal Disease Association, and the International Endotoxin Society. He is the author of over 70 scientific articles and has presented his research findings at numerous national and international meetings. He holds a U.S. patent and numerous foreign patents for his work on the application of the *Limulus* amoebocyte lysate test to evaluate gonorrhea.

Dr. Prior's research interests include the clinical applications of the *Limulus* amoebocyte lysate test, rapid diagnosis of infectious diseases using antigen detection methods, and the application of engineering principles and practices to solve medical problems (biomedical engineering). He has recently established a hemodialysis water testing laboratory in Columbus, and is a data management consultant for the AIDS Clinical Trial Unit at The Ohio State University. In addition to his research activities, he is very active as a teacher and advisor to the Infectious Disease Fellows and graduate students. He is guest lecturer at the School of Allied Medical Professions at The Ohio State University and gives lectures at community hospitals and Veteran Administration facilities in Ohio.

CONTRIBUTORS

Eduardo C. Alfonso, M.D.
Assistant Professor
Department of Ophthalmology
Bascom Palmer Eye Institute
University of Miami School of Medicine
Miami, Florida

James F. Cooper, Pharm. D.
Clinical Professor of Pharmacy
Medical University of South Carolina, and
President
Endosafe, Inc.
Charleston, South Carolina

Thomas F. DeMaria, Ph.D.
Associate Professor of Otolaryngology
Department of Otolaryngology
Otologic Research Laboratories
The Ohio State University
Columbus, Ohio

Ronald J. Elin, M.D., Ph.D.
Chief, Clinical Pathology Department
Clinical Center, National Institutes of Health
Bethesda, Maryland

J. Thomas Hjelle, Ph.D.
Associate Professor of Pharmacology
Department of Basic Sciences
College of Medicine at Peoria
University of Illinois
Peoria, Illinois

H. Donald Hochstein, Dr.P.H.
Director, Biological Testing Laboratory
Division of Product Quality Control
Center for Drugs and Biologics
Food and Drug Administration
Bethesda, Maryland

Jack Levin, M.D.
Professor of Laboratory Medicine and
Professor of Medicine
University of California School of Medicine, and
Chief, Hematology Laboratory and Blood Bank
Veterans Administration Medical Center
San Francisco, California

David J. Lim, M.D.
Professor of Otolaryngology
Department of Otolaryngology
Otologic Research Laboratories
The Ohio State University
Columbus, Ohio

Emil M. Mikolajcik, Ph.D.
Professor Emeritus of Food Science and Technology
Department of Food Science and Technology
The Ohio State University
Columbus, Ohio

Darlene Miller, M.A., M.T.(ASCP)
Supervisor of Microbiology
Bascom Palmer Eye Institute
University of Miami School of Medicine
Miami, Florida

Marcia A. Miller, Ph.D.
Professor of Microbiology
Department of Basic Sciences
College of Medicine at Peoria
University of Illinois
Peoria, Illinois

Ronald Nachum, Ph.D.
Associate Professor of Microbiology and Pathology
Department of Pathology
King/Drew Medical Center
Los Angeles, California

Frederick C. Pearson III, Ph.D.
Vice President, Scientific Operations
Cell Technology, Inc.
Boulder, Colorado

Richard B. Prior, Ph.D.
Associate Professor of Internal Medicine
Division of Infectious Diseases
Department of Internal Medicine
The Ohio State University College of Medicine, and
Associate Professor of Microbiology
The Ohio State University College of Biological Sciences
Columbus, Ohio

Edward R. Richter, Ph.D.
Adjunct Professor
Food Science and Technology
Department of Food Science and Technology
The Ohio State University
Columbus, Ohio

Carl N. Shuster, Jr., Ph.D.
Adjunct Professor of Biological Oceanography
The College of William and Mary
Virginia Institute of Marine Science
Gloucester Point, Virginia

David L. Smalley, Ph.D.
Assistant Professor of Pathology
University of Tennessee at Memphis
Memphis, Tennessee

TABLE OF CONTENTS

Chapter 1
Endotoxins and Their Clinical Significance — An Overview 1
Marcia A. Miller and J. Thomas Hjelle

Chapter 2
The American Horseshoe Crab, *Limulus polyphemus* 15
Carl N. Shuster, Jr.

Chapter 3
The *Limulus* Amoebocyte Lysate Test 27
Richard B. Prior

Chapter 4
Role of the FDA in Regulating the *Limulus* Amoebocyte Lysate Test 37
H. Donald Hochstein

Chapter 5
Detection of Endotoxemia 51
Frederick C. Pearson III

Chapter 6
Detection of Gram-Negative Meningitis 67
Ronald Nachum

Chapter 7
Detection of Gram-Negative Bacteriuria 81
Ronald Nachum

Chapter 8
Evaluation of Gonorrhea 97
Richard B. Prior

Chapter 9
Evaluation of Pyogenic Arthritis 111
Ronald J. Elin

Chapter 10
Evaluation of Otitis Media 115
Thomas F. DeMaria and David J. Lim

Chapter 11
Detection of Ocular Infections 121
Eduardo C. Alfonso and Darlene Miller

Chapter 12
Detection of Peritonitis and Perforation in Blunt Abdominal Trauma 135
David L. Smalley

Chapter 13
Allied Medical Applications of the *Limulus* Amoebocyte Lysate Test 139
Frederick C. Pearson III

Chapter 14
Veterinary Applications — An Overview 155
Edward R. Richter and Emil M. Mikolajcik

Chapter 15
Evaluation of Sterile Pharmaceuticals and Medical Devices — An Overview 159
James F. Cooper

Chapter 16
Perspectives of the *Limulus* Amoebocyte Lysate Test in Clinical Medicine 167
Richard B. Prior

Index 173

Chapter 1

ENDOTOXINS AND THEIR CLINICAL SIGNIFICANCE — AN OVERVIEW

Marcia A. Miller and J. Thomas Hjelle

TABLE OF CONTENTS

I. Introduction 2

II. Chemical Structure of Lipopolysaccharides 4
A. O-Antigen 5
B. Core Polysaccharide 5
C. Lipid A 6

III. Beneficial Effects of Endotoxin 7

IV. Detrimental Effects of Endotoxin 9

V. The LAL Test and Endotoxins 10

References 11

"You never know what is enough unless you know what is more than enough." William Blake, Proverbs of Hell

"Their most universal quality is diversity." Michel Eyquem de Montaigne in ESSAYS, *Book I (1580), To the Reader*

I. INTRODUCTION

After more than 100 years of research into the biology and chemistry of endotoxin[1] we are confronted by the reality that endotoxin is not a single molecular species but a "brew" of lipopolysaccharide (LPS), protein, and other trace components in variable amounts. Thus, generalization about the pharmacology of endotoxin preparations remains problematic. As discussed later in this chapter, the composition of LPS or endotoxin preparations used in research are probably as diverse as the number of studies conducted. The definition of endotoxin pharmacology is further complicated by the diverse biological effects induced by endotoxin in mammalian systems; these activities may be the result of only one component of the endotoxin "brew" acting on multiple pathways or an array of components acting on multiple pathways. It is important to note that the relationship between *in vitro* preparations of endotoxin and the form(s) in the host which elicits particular biological responses has yet to be clearly defined. It is against this background of endotoxin "brews" and a mammalian biology that responds to endotoxin in diverse and profound ways that we discuss current knowledge of the biology and chemistry of endotoxin.

Endotoxin is one of the two most common of all naturally occurring molecules to which humans are exposed; the other, peptidoglycan, is also of bacterial origin.[2,3] Introduction of the term "endotoxin" has been attributed to a pupil of Robert Koch, Richard Pfeiffer, in 1892. This term was actually preceded by the word "pyrogen" in 1876;[4] it is still frequently employed to describe the first and most basic of biological properties of endotoxin, i.e., fever production by endotoxin-induced endogenous pyrogen.

Because minute quantities stimulate host defense mechanisms, endotoxin has been proposed to be a life-important "exohormone".[4] However, endotoxin in somewhat larger amounts can induce endotoxic shock and even death.[5] It is this unregulated (overstimulated) response of host defense mechanisms, associated with larger amounts of endotoxin, which may actually become the driving force behind the primary/underlying disease or destructive biological pathology. Catastrophe results following endotoxin exposure, not because of any one singular event, but rather because endotoxin simultaneously induces a cascade of interactive and possibly synergistic biological events.[6] Therefore, the modification or elimination of only a single event in the cascade of biological responses to endotoxin is unlikely to significantly abrogate the deleterious effects of endotoxin as a whole.[7]

As the first quotation suggests, when applied to endotoxin, there exists a narrow but ill-defined threshold for man's response to endotoxin. The limited amount of endotoxin that imparts beneficial (protective) effects is the "enough"; beyond this level detrimental effects occur, which is the "more than enough". The quantitative effect of endotoxin on the biological responses elicited is illustrated in Figure 1. The extent to which the boundaries of beneficial and detrimental effects overlap is dependent upon modifiers such as dose, route of exposure, and exposure history of the host.[8,9] A fourth and critical, but virtually unstudied modifier is the type of endotoxin involved. Because of demonstrated heterogeneity of biological activity in preparations of the endotoxin molecule,[10] the primary variable that influences responses following exposure is the method of extraction and purification.[11] It is this difference in the chemical composition of various endotoxin preparations and the resultant effects this may have on host responses that requires more attention if we are to clearly elucidate the biological

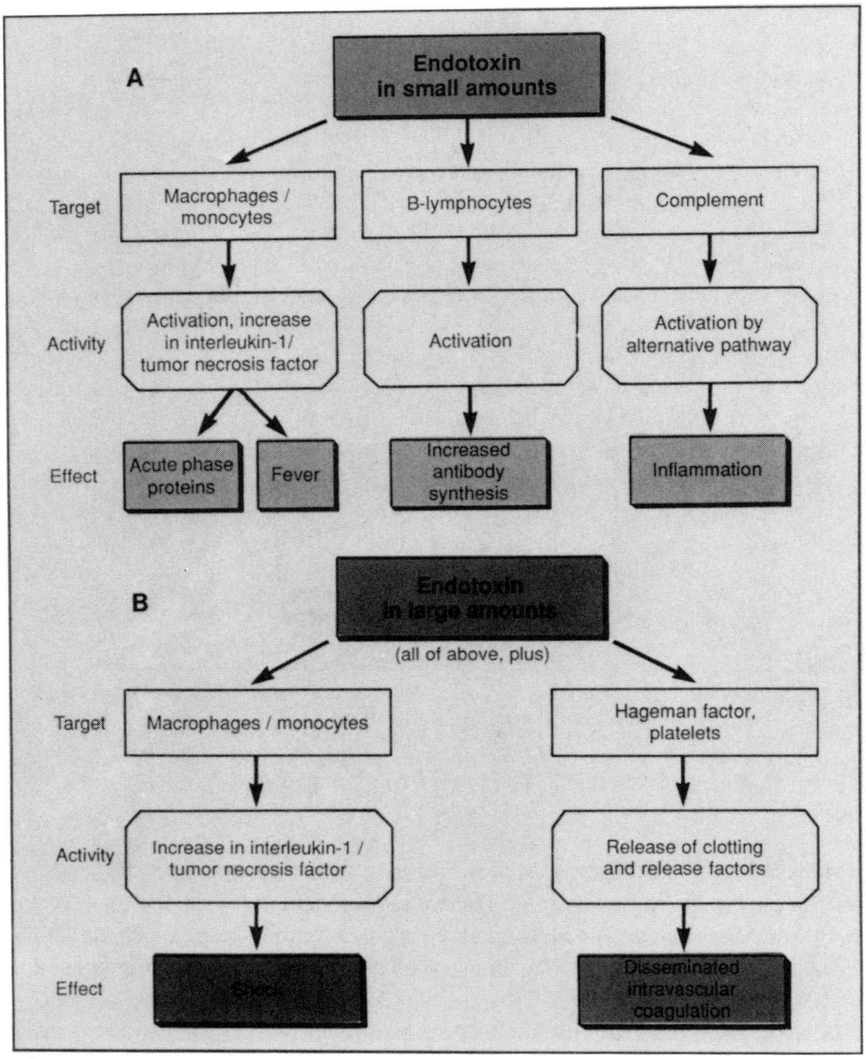

FIGURE 1. The concentration (dependent) effect on the biological activity of endotoxin. (From *Mechanisms of Microbial Disease,* Schaechter, M., Medoff, G., and Schlessinger, D., Eds., Williams & Wilkins, Baltimore, 1989. With permission.)

spectrum of endotoxin activity. The opening quote by de Montaigne very eloquently summarizes the biological nature of endotoxin. Perhaps even more important, little if any consideration has been given to how *in vitro* extraction-derived endotoxin differs chemically and in terms of activity from *in vivo* host-derived (processed) endotoxin.

The myriad of complex pathophysiological effects of endotoxin came into prominent recognition from animal studies during the 1950s and 1960s. These studies centered around attempts to circumvent the morbidity and mortality associated with septic shock in humans.[12] Endotoxin induces an enormous spectrum of host responses.[7] Of the species examined to date, humans are the most sensitive to endotoxin challenge.[3] Concentrations as low as 1.0 ng/ml may be lethal for humans.[13]

The study of the pathophysiology of endotoxemia is complex because endotoxin is not itself an effector molecule, but rather induces the release of an interrelated array of physiologic mediators.[14] The chemical and biological complexity of endotoxin holds something of interest

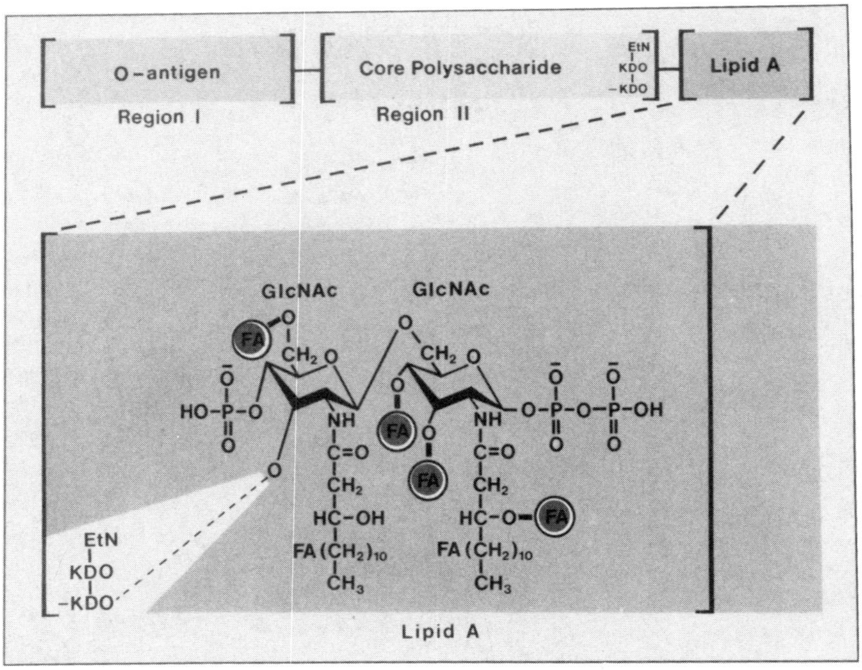

FIGURE 2. The structure of lipopolysaccharide. KDO=keto-deoxyoctanoate; FA = fatty acid; EtN = ethanolamine; GlcNac = *N*-acetyl-D-glucosamine. (From *Mechanisms of Microbial Disease*, Schaechter, M., Medoff, G., and Schlessinger, D., Eds., Williams & Wilkins, Baltimore, 1989. With permission.)

for almost every biological scientist, be they chemists, microbiologists, cell biologists, or even surgeons.[8] There are probably few other research areas that intertwine the basic scientist and clinician into such an inescapable investigative partnership.[12] Despite the phenomenon of such vast interest and research, the molecular-biological basis for many of the properties of endotoxin remain elusive.

Studies on the biological activities of endotoxin have been greatly advanced in recent years by two important contributions: (1) use of the endotoxin hypo-responsive mouse (C3H/HeJ)[11] and (2) use of synthetic and ultrapure preparations of endotoxin.[15] The chemical synthesis of lipid A was achieved in 1985 by Imoto et al.[16] Various analogs of lipid A and partial structure have been prepared. Subtle changes in structure have been shown to modify such biological activity as pyrogenicity, lethality, and Schwartzman reactivity.[17,18] These compounds have revealed important information in elucidating structure-function relationships of the endotoxin molecule.[19] Perhaps even more fundamental and equally intriguing to man's response to endotoxin is the evolutionary significance of this toxic cornerstone of the Gram-negative cell wall.

The remainder of this chapter discusses the chemical structure of LPSs and their beneficial and detrimental effects on humans.

II. CHEMICAL STRUCTURE OF LIPOPOLYSACCHARIDES

The toxic portion of Gram-negative bacterial cell envelopes resides in the LPS component and is called endotoxin. LPS is a unique and complex molecule that is not found elsewhere in nature. The molecular weight of LPS is 400,000 to 4 million D;[20] that of free lipid A is 2000.[4] It is found only in the outer leaflet of the outer membrane bilayer of Gram-negative bacteria,

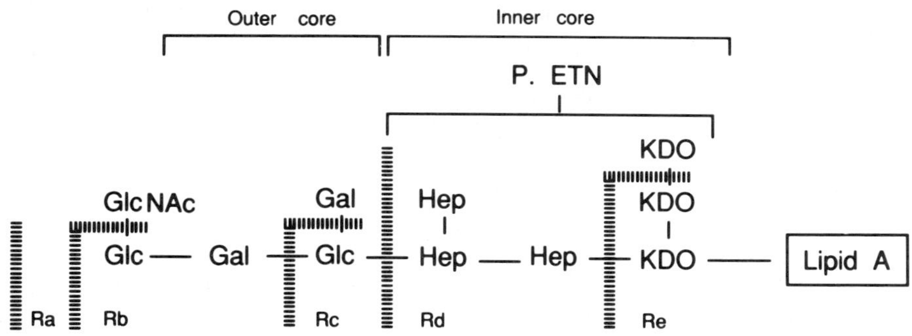

FIGURE 3. Diagrammatic representation of the core oligosaccharide and various core mutants (Ra - Re) from *Salmonella minnesota*, designated by EtN = ethanolamine; P = phosphate; Glc = D-glucose; Gal = D-galactose; GlcNac = *N*-acetyl-D-glucosamine. (From Brade, H., Brade, L., and Rietschel, E. Th., *Zentralbl. Bakteriol. Hyg.*, 268, 151, 1988. With permission.)

making the membrane asymmetric.[21] This asymmetry allows for specific recognition by the immune system and interaction with antibiotics and bacteriocins, and confers net charge for cell to cell repulsion. Although endotoxin was originally thought to be released only when cells lysed, it is now known that many Gram-negative bacteria continuously "shed" LPS[22] throughout their life cycles, losing amounts approaching 5% of total cellular endotoxin per generation.[23] Tesh and Morrison[24] have reported amounts as high as 30% under conditions that more closely reflect *in vivo* situations.

LPS consists of three distinct but covalently linked regions. The three regions (Figure 2), starting with the outermost, are (1) O-specific polysaccharide side-chains (O-antigen), (2) core polysaccharide, and (3) lipid A.

A. O-ANTIGEN

This region is probably the best known part of the LPS molecule. The O-specific chains are a terminal series of repeating[23] oligosaccharide combinations found on the outermost wall; they make up the type-specific haptenic determinants.[25] These determinants give rise to specific O-antibodies and have been employed for decades in the classification and taxonomy of serotypes of the Enterobacteriaceae, as exemplified by the Kauffmann-White classification of Salmonellae. The O-antigen is commonly known as the somatic antigen; it is also known as the "smooth" antigen due to the smooth surface of the colonies conferred by the O-polysaccharide side chains. The structure and composition of this region of the LPS molecule demonstrate great diversity, as is exhibited by some 1500 serotypes of Salmonella. Lack of the O-chains, due to defective LPS biosynthesis, results in specific rough (R) mutants, known as Ra mutants,[26] of the core polysaccharide. In addition to their "rough" surface on colonies, these mutants also demonstrate increased susceptibility to many antibiotics.

B. CORE POLYSACCHARIDE

The core region displays little heterogeneity and is very similar within a given genus. The common core polysaccharide is also known as the "rough" colony antigen.[25] The major physiological functions of the core are translocation and integration of LPS into the outer membrane and conference of stability to the outer membrane via cation linkages between proteins and LPS.[28] This region can be subdivided into (1) the lipid-A-proximal inner core and (2) the lipid-A-distal outer core (Figure 3).[19] The inner core region contains two obligatory constituents of unusual sugars unique to bacteria. These are the eight carbon 2-keto-3-deoxyoctonic acid (KDO) and a seven-carbon heptose. The reducing group KDO comprises the direct link to lipid A. Synthetic production of free lipid A in recent years has revealed the importance of the KDO linkage and the influence it exerts on the ultimate potency of endotoxin biological activity.[4] It is important to note that both KDO and heptose have been found to be lacking in certain LPS molecules, such as those in *Bacteroides fragilis*.[28] The net effect is loss of many of the classical biological activities of *B. fragilis* endotoxin. Other core mutants have also been identified, as depicted in Figure 3.

C. LIPID A

Compared to the O-chains and the core polysaccharide, lipid A is the most conserved part of the LPS molecule. Lipid A is similar to glycerol-based phospholipids with molecules of fatty acids attached to the sugars (Figure 2); one fatty acid, 3-hydroxy myristic acid, is an unusual C14 fatty acid unique to the LPS molecule.[29] The fatty acid moieties impart hydrophobicity to the molecule which serves to anchor LPS to the outer membrane. Lipid A is linked via glycoside bonds to peptidoglycan. Although lipid A from different genera may show some variability or, microheterogeneity,[30] the common element to date is the lipid A backbone, diphosphorylated β 1,6-D-glucosamine disaccharide. The common lipid A backbone is absent in at least two species of *Pseudomonas*, *P. diminuta* and *P. vesicularis*;[31] neither species exhibits endotoxic activity as measured by the *Limulus* amoebocyte lysate (LAL) assay.

These differences in the structure of endotoxin and resulting presence/absence of classical biological activity are important concepts in subsequent chapters where use of the LAL assay for detecting endotoxin is reviewed. Functional lipid A,[7] i.e., the common lipid A backbone, is a requirement of the LAL assay. Further, any test for endotoxin must be sensitive enough to detect levels, usually a few picograms of endotoxin, significant for humans.[4]

Although there is abundant evidence indicating that the primary toxicity of bacterial endotoxin resides in the lipid A component of LPS, recent studies have indicated multiple activities for both protein-associated and polysaccharide components of endotoxin.[32] Lipid A associated protein (LAP) and KDO have been shown to induce mitogenic responsiveness.[11,33] Vukajlovich et al.[34] have reported that polysaccharide can also initiate activation of the alternate pathway of human complement. The double long-chain fatty acid structures attached to lipid A are necessary for pyrogenicity and Schwartzman activity.[4]

The classic and perhaps most widely applicable procedure for extraction of smooth LPS is the Westphal et al.[35,36] method, which uses hot phenol-water for separating out LPS. Another common procedure is extraction by aqueous butanol solutions, as described by Morrison and Leive.[37] In the first procedure, LAP is dissociated from lipid A, while it is retained in the second procedure. These methods serve as examples of how differences in extraction procedures can result in differing end products. It is significant that the endotoxin obtained by aqueous butanol extraction is mitogenic in the C3H/HeJ model, while the Westphal et al. derived product is not. Because diverse chemical products can be obtained via various extraction procedures and can demonstrate differing biological activity, it would not be unexpected if yet another alteration(s) of endotoxin was found to occur during *in vivo* processing (metabolism, binding).

The validity of presumed correlation of responses from chemically prepared endotoxin with those that occur from *in vivo* processed endotoxin in humans is being challenged. Several investigators have reported release of Gram-negative membrane components upon exposure of bacteria to human serum.[38,39] A recent paper by Tesh and Morrison[40] has demonstrated significant differences in the biological activity of serum-associated LPS (s-LPS) compared to the high molecular weight aggregates of phenol-water extracted endotoxin. They reported diminished pathophysiologic effects of s-LPS, such as decreased LAL activity and lethality, but showed augmented immunostimulatory capabilities, i.e., increased mitogenicity.

Because of the diversity of components which influence the overall biological activity of endotoxin, it should become apparent that a discussion of the individual beneficial and detrimental effects will be separated, at best, by an arbitrary boundary. For example, immunogenicity is beneficial unless it progresses to hypersensitivity or other pathological consequences.[5] There is no linear scale on which good and bad responses to endotoxin are clearly defined or resolved; there appears to be an inseparable overlap of functional effects. To quote from Ivan Bennett's *Approaches to the Mechanisms of Endotoxin Action*,[41] "(endotoxins) seem to have been endowed by Nature with virtues and vices in the exact and glamorous proportions needed to render them irresistible...".

If the functional dichotomy and chemical diversity of endotoxin is acknowledged, the reader will be better able to put into perspective the clinical importance of endotoxin and the significance of a positive or negative LAL assay. It is important to note, however, that many of the responses induced by endotoxin can also be elicited by a variety of other microbial products, including those from Gram-positive bacteria, such as lipoteichoic acid (LTA). LTA is "shed" from viable cells. Like LPS, it exhibits some of the same biological effects, such as inducement of Schwartzman lesions and fixing of complement.[22] Until we fully understand the mechanisms that "permit" any response to endotoxin to exert both beneficial and toxic effects, it will not be possible to capitalize, for the benefit of humankind on its potential beneficial effects while limiting its toxicity.

III. BENEFICIAL EFFECTS OF ENDOTOXIN

A discussion of the beneficial effects of endotoxin is affected by (1) lack of emphasis in the literature and (2) status of the host. In the spectrum of publications that have been written on endotoxin, beneficial effects have not, for the most part, been particularly emphasized when compared with toxic effects. A recent book devoted to beneficial effects has been written by Nowotny.[5] In reality, any distinction between beneficial and detrimental may be dependent upon the context or framework in which the response is to be used.[42] The following paragraphs briefly discuss some of the beneficial effects with the caveat that the amount of endotoxin present under any given condition could tip the balance toward a detrimental effect.

LPS can interact directly with components of the immune system. Those effects of endotoxin considered to be beneficial include, but not are limited to, immunogenicity, immunomodulation (mitogenicity and immune adjuvant effect), nonspecific resistance, tolerance, and induction of colony stimulating factor (CSF) and γ-interferon. In addition to direct interaction with components of the immune system, LPS can also induce an array of cytokines which mediate pathophysiologic responses of Gram-negative shock. Various beneficial and detrimental responses to endotoxin exposure have also been attributed to cytokines. Cytokines are biologically active mediators produced by cells of the reticuloendothelial system (RES), such as macrophages; these cells can be stimulated by whole microbes or their components, such as endotoxin. Two examples of cytokines with apparent beneficial effects are CSF and γ-interferon.

Humoral immune responses can be elicited by each of the three regions of the LPS molecule, in addition to LAP (endotoxin protein).[43] This is of primary importance in developing schemes for protecting against endotoxin. The O-antigens are of such diversity as to preclude their use in developing protective antisera. Lipid A also appears to be an unlikely immunizing candidate because (1) the antigen is highly insoluble in water and (2) its membrane location is deeply buried in the outer bacterial membrane, making it unlikely to be accessible to antibodies.[44] Studies have demonstrated that the core polysaccharide has potential as an antigen for developing protective antisera.[45-47] In a study of human patients by Ziegler and colleagues,[46] involving the use of hyperimmune serum to "core" polysaccharide, mortality from Gram-negative bacteremia was cut in half. If shock did occur, the probability of recovery was 2.8 times greater than in patients exhibiting normal serum. Interestingly, Gaffin and associates have observed that blood units from blood banks can contain high concentrations of antibodies to endotoxin and thus could be a ready biological source of anti-endotoxin. Significantly, the antibodies were found to bind to several Gram-negative bacterial strains.[48,49] Although these findings are promising, the extent to which immune responses to the LPS molecule may actually influence resistance or susceptibilty to Gram-negative infections remains unresolved.

Endotoxin has also been demonstrated to exert immunomodulatory effects, including mitogenicity and adjuvancy. These properties have the potential to be at once beneficial and harmful. They are included under beneficial effects due to their potential role in the immunotherapeutic potential of LPS for tempering immune responses and affecting immunity to this molecule in the treatment of Gram-negative sepsis.[10]

The C3H/HeJ endotoxin hyporesponsive mouse has played an invaluable role in elucidating the mitogenic factor in LPS.[50,51] The mitogenic response is carried out only by B cells;[1] LPS does not directly activate T cells but has been found to stimulate T-cell differentiation.[52] Lymphocyte activation by LPS has been found to be caused by LAP or the endotoxin protein.[53] Characteristic toxic effects of endotoxin, such as hypotension and disseminated intravascular coagulation (DIC), are not reported for LAP.[5] The therapeutic role of this mitogen in functioning as a specific immunogen, activator of lymphocytes, and/or inducer of important cytokines, awaits further studies.

The immune adjuvant effect of endotoxin was discovered in 1956 by Johnson et al.[5] Adjuvancy has been stated to be one of the potentially most beneficial effects of endotoxin, especially as it applies to endotoxin as an adjuvant in vaccines. Adjuvancy of endotoxin has been demonstrated in a number of animal models by showing increases in (1) accelerated antigen eliminations, (2) persistence of circulating antibodies, (3) increases in the number of antibodies, often to unrelated antigens,[54] and (4) a shortening of the lag period in the immune response.[1]

Early on, endotoxin tolerance and induction of nonspecific resistance (NSR) appeared to be antibody independent, occur after the same time interval following endotoxin exposure, and have an underlying basis of nonspecific activation of phagocytic cells.[55] NSR, first discovered in 1956,[56] can be demonstrated against infection caused by a variety of microorganisms including fungi, parasites, viruses, and both Gram-negative and positive organisms. The balance between NSR and susceptibility to infection is a precarious one, however, and is dependent upon dose, route of administration, and the time interval between endotoxin exposure and challenge with an infectious agent.[56-58] Because of the narrow threshold between beneficial and toxic effects of endotoxin, the use of endotoxin for induction of NSR is precluded. Attempts have been to "detoxify" endotoxin while preserving its beneficial potential. In this regard, investigators have employed a variety of procedures, using principles derived from the preparation of tetanus and diphtheria antitoxins;[59] radiation detoxified LPS is an example.[60] Irradiated preparations have shown a reduction and/or loss of toxic effects of endotoxin. They have, however, retained many beneficial effects by triggering host defenses in a nonspecific immunologically manner, such as through promotion of phagocytosis and RES activation.[61]

The refractory state of endotoxin tolerance is a long-observed phenomenon, the basic concept of which was unwittingly recognized almost 100 years ago by physicians employing bacterial vaccines for fever therapy.[62] Tolerance surprisingly is easy to induce and thus may play an important role in naturally occurring disease.[41] The two most extensively studied examples of tolerance are fever production (pyrogenicity) and lethality of endotoxin. Both immunologic and nonimmunologic mechanisms play a role in mediating tolerance; the RES macrophage represents a major target for both endogenous pyrogenicity and lethality.[62] This is not surprising, as the RES is very important in the metabolism of LPS, especially the Kupffer cells of the liver.[63] There does not appear to be a single mechanism underlying the many tolerances to the diverse responses to endotoxin. Tolerance is a very complex phenomenon, perhaps as complex as the chemistry and biological activity of endotoxin. As with many other responses elicited by endotoxin, tolerance is influenced or affected by various modifiers such as dose and antigenicity. Studies by an array of investigators have demonstrated that tolerance to LPS of certain bacterial pathogens (i.e., *Salmonella*) does not necessarily alter acquisition of the actual disease (e.g., typhoid fever); LPS from some organisms actually produces hypersensitivity (*Brucella* sp.).[64-68] Its relevance to resistance to or recovery from human disease remains unclear.

Recent studies by Vogel and associates[69] have indicated that CSF plays an important role in the development of endotoxin tolerance. Because of this, the function of CSF appears to be a defensive one primarily directed against endotoxin. Lipid A has been demonstrated to be the active component stimulating CSF generation.[70] CSF is actually a family of glycoproteins that share the common activities of (1) promoting proliferation and differentiation of bone marrow-derived precursor cells and (2) modifying the function of mature cells through membrane changes, modulation of receptor expression, synthesis of prostaglandins, and stimulation of phagocytosis.[7,71] The original work on this stimulatory glycoprotein substance was conducted in the early 1960s by Pluznik and Sachs[72] and Bradley and Metcalf.[73]

Like tolerance, the production of γ-interferon is another example of an event resulting from a series of complex interrelated responses to endotoxin. Endotoxin can induce interferon production in animals[74-76] and in humans.[77] The essential component for interferon induction is lipid-A;[78] its induction is not direct, but depends upon interleukin-2 (IL-2).[79] γ-Interferon mediates a variety of important host responses carried out by the macrophage, such as enhancement of oxidative metabolism and stimulation of phagocytic activity.[7] In contrast, γ-interferon has been proposed to be another endogenous pyrogen.[74] A recent study using human subjects was unable to demonstrate detectable levels of γ-interferon following intravenous (i.v.) administration of endotoxin, although tumor necrosis factor (TNF) was readily detectable.[80] It remains to be determined whether lack of corroborating data indicate a diminishing role of interferon in the response to endotoxin and/or Gram-negative infections or is simply due to limitations of the assays used. Whatever the ultimate role of this immunoregulatory cytokine, the probability of developing a nontoxic interferon inducer from endotoxin appears unlikely at the present time.[74]

IV. DETRIMENTAL EFFECTS OF ENDOTOXIN

Endotoxin adversely affects almost every organ system of the body and many metabolic pathways. For example, the anterior hypothalamus of the central nervous system produces fever in response to endogenous pyrogen produced by various phagocytic cells, which in turn are induced by endotoxin.[81,82] The cardiovascular system and LPS interact in a complex fashion, resulting in the syndrome of endotoxic shock.[83] The hemopoietic system adversely responds to endotoxin with the manifestation of DIC.[84] Although DIC is a concomitant feature of endotoxic shock, a determination as to its role in endotoxic shock has yet to be made.[1] Immunosuppressant activity may be yet another example of detrimental overtones for endotoxin. Endotoxin affects carbohydrate metabolism via interleukin-6 (IL-6) by downregulating the gluconeogenic enzyme phosphoenolpyruvate carboxykinase (PEPCK), which depresses blood sugar to hypoglycemic levels. Endotoxin-induced hypoglycemia has been demonstrated to be of such a magnitude as to result in death in animals.[85] The hypotension observed in endotoxic shock is, in part, due to the recently implicated endogenous opioids, the endorphins.[1]

Although TNF is apparently beneficial because of its tumoricidal activity by human monocytes, it is an important cause of immunopathology. TNF is produced by the endotoxin/LPS induced activation of primed macrophages;[86] responses to TNF are effected through the cyclooxygenase pathway.[80] It is likely that TNF is synonymous with the modulator protein cachectin.[87] This cytokine is considered to be a major mediator of the circulatory collapse and tissue necrosis observed in Gram-negative septicemia. The harmful effects of TNF are focused on endothelial cells and neutrophils.[88] The anticoagulant property of endothelial cells is lost and fibrin deposition is enhanced, which can promote DIC when endothelial cells are challenged with TNF. Equally important is the contraction of vascular endothelial cells, which results in fluid loss from the circulation into tissue spaces, causing edema and hypovolemia.[89] TNF also promotes the adherence of neutrophils to the endothelium, resulting in degranulation of neutrophils with release of tissue damaging oxygen intermediates, such as the superoxide anion and hydrogen peroxide. TNF can induce interleukin-l (IL-1) and shares the intrinsic property of pyrogenicity with IL-1. Michie et al.[80] report that TNF is detectable in humans only at the onset of clinical responses, but not when responses are at a maximum. Perhaps this is the reason why other endotoxin-induced cytokines have not been readily detected in patients.

Platelet-activating factor (PAF) is stimulated by endotoxin; like TNF it is a mediator of circulatory collapse. PAF is not a single molecular species but a family of lipid mediators. It affects many of the same cells as TNF and produces similar diverse biological activities. There may be a synergistic effect between endotoxin and PAF in producing certain endotoxin-associated pathology such as bowel necrosis.[90] It is speculated that the pathology of bowel necrosis is augmented by the endotoxin-induced secondary mediators, prostaglandins, and leukotrienes. These mediators are produced from arachidonic acid metabolism via the cyclooxygenase and lipoxygenase pathways.[7,91] It is of interest that PAF is also induced by IL-1. What is emerging, therefore, is another example of the complex interaction of endotoxin and primary-secondary mediators resulting in a myriad of pathophysiological responses.

Although IL-1 may be viewed as more physiologically beneficial[92,93] than harmful, complex interreactions with other detrimental mediators necessitates their mention. As with the other previously discussed mediators, IL-1 represents a family of molecules (cytokines). Their biological activities are about as diverse as endotoxin itself.[7] Potent vasoactive compounds, such as PAF and the arachadonic acid metabolites, prostaglandins and leukotrienes, are also stimulated by IL-1. Thus, the list of mediators known to play critical interactive roles in the host response to endotoxin is large and constantly expanding.

Induction of tissue factor procoagulant activity is considered to play an important part in coagulopathies associated with endotoxemia.[7] Diffuse microvascular thrombosis with resultant loss of cellular perfusion due to occlusion is considered to be a factor involved in multiple organ failure and death observed with endotoxic shock.[92] Evidence exists implicating both the intrinsic (via activation of factor XII) and extrinsic coagulation pathways, as influenced by an endotoxin-induced expression of an endogenous (surface) glycoprotein called tissue factor.[7] An active protease is formed when tissue factor and blood factor VII bind, resulting in the cleavage of factors X and IX. This cleavage results in the extrinsic coagulation cascade and fibrin deposition.[7] Tissue factor is inhibited by corticosteroids, thus serving as a likely basis for their use in managing endotoxic shock; TNF is also inhibited by corticosteroids.[94]

Because of the dual beneficial and detrimental roles of endotoxin, the final analysis of endotoxin may reveal that it is an important promoter of both health and disease in humans.[41]

V. THE LAL TEST AND ENDOTOXINS

As noted, endotoxins are complex substances associated with Gram-negative bacteria. Their role in the disease process is complicated and their detection is important in evaluating and treating infectious processes involving these bacteria. The many uses of the LAL test discussed in this book demonstrate the importance of this test in clinical medicine; the application of the LAL test in both biomedical research and clinical medicine remains unlimited. It is especially applicable in elucidating conditions where responses to endotoxin resemble other disorders, ranging from hypersensitivity states to dietary injuries.[41] In contrast, endotoxin may be an elusive initiator or exacerbator of disease, such as autosomal dominant polycystic kidney disease (PKD). Endotoxin has been demonstrated to be procystogenic in animals,[95] and Miller et al.[96,97] have recently reported that PKD patients, asymptomatic for urinary tract infections, exhibited significant levels of detectable endotoxin in their urine (endotoxiuria) compared with normal controls.

Finally, in some disease states, endotoxin (the lipid A moiety) cannot always be demonstrated using the LAL test. The most notable of these conditions is Gram-negative sepsis.[98,99] This raises the important question of which biologically active component(s) of the LPS molecule might be present, yet undetected, and playing an important role in the clinical syndrome of sepsis. As tests become routinely available for detecting important mediators (i.e., TNF and IL-1-β,[100,101]) of sepsis, endotoxic shock, and other Gram-negative infections, the LAL test will likely become part of an important battery of tests for diagnosing, managing, and treating these complex clinical syndromes.

REFERENCES

1. **Elin, R. J. and Wolff, S. M.,** Biology of endotoxin, *Annu. Rev. Med.*, 27, 127, 1976.
2. **Rose, W.,** Infections of bacterial toxins in the toxicity of chemotherapeutic agents, *Crit. Rev. Toxicol.*, 159, 1973.
3. **Chetty, C. and Schwab, J. H.,** Endotoxin-like products of Gram-positive bacteria, in *Handbook of Endotoxins*, Vol. 1, *Chemistry of Endotoxin*, Rietschel, E. Th., Ed., Elsevier, Amsterdam, 1984, 376.
4. **Westphal, O.,** Introduction, in *Handbook of Endotoxin*, Vol.1, *Chemistry of Endotoxin*, Rietschel, E. Th., Ed., Elsevier, Amsterdam, 1984, xvii.
5. **Nowotny, A.,** In search of the active site in endotoxins, in *Beneficial Effects of Endotoxins*, Nowotny, A., Ed., Plenum Press, New York, 1983, 1.
6. **Thomas, L.,** An apology, in *The Medusa and the Snail*, Viking, New York, 1979, 88.
7. **Morrision, D. C. and Ryan, J. L.,** Endotoxins and disease mechanisms, *Rev. Med.*, 38, 417, 1987.
8. **Berry, L. J.,** Introduction, in *Handbook of Endotoxin*, Vol. 3, *Cellular Biology of Endotoxin*, Berry, L. J. Ed., Elsevier, Amsterdam, 1985, xvii.
9. **Schlessinger, D.,** Biology of infectious agents, in *Mechanisms of Microbial Disease*, Schaechter, M., Medoff, G., and Schlessinger, D., Eds., Williams & Wilkins, Baltimore, 1989, 17.
10. **Jacobs, D. M.,** Immunomodulatory effects of endotoxin, in *Beneficial Effects of Endotoxin*, Nowotny, A., Ed., Plenum Press, New York, 1983, 307.
11. **Morrison, D. C.,** Bacterial endotoxins and pathogenesis, *Rev. Infect. Dis.*, 5(Suppl. 4), S733, 1983.
12. **Hinshaw, L. B.,** Introduction, in *Handbook of Endotoxin*, Vol. 2, *Pathophysiology of Endotoxin*, Hinshaw, L. B., Ed., Elsevier, Amsterdam, 1985, xix.
13. **Wardle, E. N.,** Endotoxin and acute renal failure, *Nephron*, 14, 321, 1975.
14. **Young, L. S.,** The complex challenge of bacteremia and endotoxemia, in *Bacterial Endotoxins: Pathophysiological Effects, Clinical Significance, and Pharmacological Control*, Alan R. Liss, New York, 1988, 209.
15. **Rietschel, E. T., Brade, H., Brade, L., Brandenburg, K., Schade, U., Seydel, U., Zahringer, U., Galanos, C., Luderitz, O., Westphal, O., Labischinski, H., Kusumoto, S., and Shiba, T.,** Lipid A, the endotoxic center of bacterial lipopolysaccharides: relation of chemical structure to biologic activity, *Prog. Clin. Biol. Res.*, 231, 25, 1987.
16. **Imoto, M., Yoshimura, H., Sakagucki, N., Kusumoto, S., and Shiba, T.,** Total synthesis of *Escherichia coli* lipid A, *Tetrahedron Lett.*, 26, 1545, 1985.
17. **Galanos, C. T., Hansen-Hagge, T., Lehmann, V., and Luderitz, O.,** Comparison of the capacity of two lipid A precursor molecules to express the local Schwartzman phenomenon, *Infect. Immun.*, 48, 355, 1985.
18. **Galanos, C., Lehmann, V., Luderitz, O., Rietschel, E. T., Westphal, O., Brade, H., Brade, L., Freudenberg, M. A., Hansen-Hagge, T., Luderitz, T., McKenzie, G., Schade, U., Strittmatter, W., Tanamoto, K., Zahringer, U., Inoto, M., Yoshimura, H., Yamamoto, M., Shimamoto, T., Kusumoto, S., and Shiba, T.,** Endotoxic properties of synthetic lipid A part structures. Comparison of bacterial free lipid A and the lipid A disaccharide precursor with chemically synthesized lipid A precursor and analogues, *Eur. J. Biochem.*, 140, 221, 1984.
19. **Brade, H., Brade, L., and Rietschel, E. Th.,** Structure-activity relationships of bacterial lipopolysaccharides (endotoxins), *Zentralbl. Bakteriol. Hyg.*, 268, 151, 1988.
20. **Brock-Utne, J. G. and Gaffin, S. L.,** Endotoxins and anti-endotoxins (their relevance to the anaesthetist and the intensive care specialist), *Anesth. Intensive Care*, 17, 49, 1989.
21. **Muhlradt, P. and Golecki, J.,** Asymmetrical distribution and artificial reorientations of lipopolysaccharide in the outer membrane bilayer of *Salmonella typhimurium*, *Eur. J. Biochem.*, 51, 343, 1975.
22. **Nowotony, A.,** Shedding bacteria, *Biomembranes*, 11, 1, 1983.
23. **Volk, W. A., Benjamin, D. C., Kadner, R. J., and Parsons, J. T.,** *Essentials of Medical Microbiology*, 3rd ed., Lippincott, Philadelphia, 1986, 238.

24. **Tesh, V. L. and Morrison, D. C.**, The interaction of *Escherichia coli* with normal human serum; factors affecting the capacity of serum to mediate lipopolysaccharide release, *Microb. Pathogen.*, 4, 175, 1988.
25. **Jaewtz, E., Melnick, J. L., Adelberg, E. A., Brooks, G. F., Butel, J. S., and Ornston, L. N.**, *Medical Microbiology*, 18th ed., Appleton & Lange, Norwalk, CT, 1989, 6.
26. **Luderitz, O. and Galanos, C.**, Endotoxins of gram-negative bacteria, in *Pharmacology of Bacterial Toxins*, Sect. 119, *International Encyclopedia of Pharmacology and Therapeutics*, Dorner, F. and Drews, J., Eds., Pergamon Press, Oxford, 1986, 307.
27. **Schindler, M. and Osborn, M. J.**, Interaction of divalent cations and polymyxin B with lipopolysaccharide, *Biochemistry*, 18, 4425, 1979.
28. **Bergan, T.**, Pathogenicity of anaerobic bacteria, *Scand. J. Gastroenterol.*, 91, 1, 1984.
29. **Luderitz, O., Jann, K., and Wheat, R.**, Somatic and capsular antigens of gram-negative bacteria, in *Comprehensive Biochemistry*, Florkin, M. and Stotz, E. H., Eds., Elsevier, Amsterdam, 1968, 105.
30. **Rietschel, E. Th., Galanos, C., Luderitz, O., and Westphal, O.**, The chemistry and biology of lipopolysaccharides and their lipid A component, *Immunopharmacology and the Regulation of Leukocyte Function*, Webb, D. R., Ed., Marcel Dekker, New York, 1982, 183.
31. **Wilkinson, S. G. and Taylor, D. P.**, Occurrence of 2,3,diamino-2,3-dideoxy-D-glucose in lipid A from lipopoly-saccharide of *Pseudomonas diminuta*, *J. Gen. Microbiol.*, 109, 367, 1978.
32. **Morrison, D. C. and Ryan, J. L.**, Endotoxin and disease mechanisms, *Annu. Rev. Med.*, 38, 417, 1987.
33. **Flebbe, L. D., Morrison, D. C., and Vukajlovich, S. W.**, Mitogenic stimulation of C3H/HeJ LPS-low responder splenocytes by R-chemotype LPS preparations, *Fed. Proc.*, 46(Abstr.), 1208, 1987.
34. **Vukajlovitch, S. W., Hoffman, J., and Morrison, D. C.**, Activation of human serum complement by bacterial lipopolysaccharides: structural requirements for antibody independent activation of the classical and alternative pathways, *Mol. Immunol.*, 24, 319, 1987.
35. **Westphal, O. and Jahn, K.**, Bacterial lipopolysaccharides: extraction with phenol-water and further applications of the procedure, in *Methods in Carbohydrate Chemistry*, Wistler, R. L., Ed., Academic Press, New York, 1965, 83.
36. **Westphal, O., Luderitz, O., and Blister, F.**, Uber die Extraktioinen von Bakterien mit Phenol-Wasser, *Z. Naturforsch. Teil B*, 7, 148, 1955.
37. **Morrison, D. and Leive, L.**, Fractions of lipopolysaccharide from *E. coli* 0111:B4 prepared by two extraction procedures, *J. Biol. Chem.*, 250, 2911, 1975.
38. **Demonty, J. and DeGraeve, J.**, Release of endotoxin lipopolysaccharide by sensitive strains of *Escherichia coli* submitted to the bactericidal action of human serum, *Med. Microbiol. Immunol.*, 170, 265, 1982.
39. **Tesh, V. L., Duncan, R. L., Jr., and Morrison, D. C.**, The interaction of *Escherichia coli* with normal human serum; the kinetics of serum-mediated lipopolysaccharide release and its dissociation from bacterial killing, *J. Immunol.*, 137, 1329, 1986.
40. **Tesh, V. L. and Morrison, D. C.**, The physical-chemical characterization and biologic activity of serum released lipipolysaccharides, *J. Immunol.*, 141, 3523, 1988.
41. **Bennett, I. L., Jr.**, Approaches to the mechanisms of endotoxin action, in *Bacterial Endotoxins*, Landy, M. and Braun, W., Eds., Quinn & Boden, Rahway, NJ, 1964, xiii.
42. **Berry, L. J. and Gaska, J. E.**, The role of humoral factors in endotoxin-induced reactions, in *Beneficial Effects of Endotoxins*, Nowotny, A., Ed., Plenum Press, New York, 1983, 355.
43. **Neter, E.**, Immunogenicity of endotoxin, in *Beneficial Effects of Endotoxins*, Nowotny, A., Ed., Plenum Press, New York, 1983, 91.
44. **Braude, A. I. and Ziegler, E. J.**, Protection against Gram-negative bacteremia with antiserum to endotoxins, in *Beneficial Effects of Endotoxins*, Nowotny, A., Ed., Plenum Press, New York, 1983, 111.
45. **Zinner, S. and McCabe, W.**, Effects of IgM and IgG antibody in patients with bacteremia due to gram-negative bacilli, *J. Infect. Dis.*, 133, 32, 1976.
46. **Ziegler, E., McCutchan, J. A., Fierer, J., Glauser, M., Sadoff, J. C., Douglas, H., and Braude, A. I.**, Treatment of gram-negative bacteremia and shock with human anti-serum to a mutant *Escherichia coli*, *N. Engl. J. Med.*, 307, 1225, 1982.
47. **Wolf, J., McCutchan, J., Ziegler, E., and Braude, A.**, Prophylactic antibody to core lipopolysaccharide in neutropenic patients, in *Proc. 11th Int. Cong. Chemother.* and *19th Intersci. Conf. Antimicrob. Agents Chemother.*, American Society for Microbiology, Washington D.C., 1979, 65.
48. **Gaffin, S. L., Badsha, N., Vorster, B., Conradie, J., and Brock-Utne, J. G.**, An ELISA procedure for detecting human anti-endotoxin antibodies in serum, *Ann. Clin. Biochem.*, 19, 191, 1982.
49. **Gaffin, S. L., Badsha, N., and Vorster, B.**, Properties of human lipopolysaccharide gamma globulin: specificity and protective effects, *Vox. Sang.*, 48, 276, 1985.
50. **Sultzer, B. M.**, Genetic control of host responses to endotoxin, *Infect. Immun.*, 5, 107, 1972.
51. **Sultzer, B. M.**, Mitogenic and immune responses to endotoxin: a deficiency in C3H/HeJ mice, *Abst. Annu. Meet. Am. Soc. Microbiol.*, 1973, 85, M69.

52. **Sultzer, B. M.,** Lymphocyte activation by endotoxin and endotoxin protein: the role of the C3H/HeJ mouse, in *Beneficial Effects of Endotoxins*, Nowotny, A., Ed., Plenum Press, New York, 1983, 227.
53. **Betz, S. J. and Morrison, D. C.,** Chemical and biological properties of a protein rich fraction of bacterial lipopolysaccharides. I. The *in vitro* murine lymphocyte response, *J. Immunol.*, 119, 1475, 1977.
54. **Johnson, A. G.,** Adjuvant action of bacterial endotoxins on antibody formation: a historical perspective, in *Beneficial Effects of Endotoxins*, Nowotny, A., Ed., Plenum Press, New York, 1983, 249.
55. **Parant, M.,** Effect of LPS on the nonspecific resistance to bacterial infections, in *Beneficial Effects of Endotoxins*, Nowotny, A., Ed., Plenum Press, New York, 1983, 179.
56. **Rowley, D.,** Rapidly induced changes in the level of nonspecific immunity in laboratory animals, *Br. J. Exp. Pathol.*, 37, 223, 1956.
57. **Landy, M. and Pillemer, L.,** Increased resistance to infection and accompanying alteration in properdin levels following administration of bacterial lipopolysaccharides, *J. Exp. Med.*, 104, 383, 1956.
58. **Dubos, R. J. and Schaedler, R. W.,** Reversible changes in the susceptibility of mice to bacterial infections, *J. Exp. Med.*, 104, 53, 1956.
59. **Sultzer, B. M.,** Chemical modification of endotoxin and inactivation of its biological properties, in *Microbial Toxins*, Vol. 5, Kadis, S., Weinbaum, G., and Ajl, S. J., Eds., Academic Press, New York, 1981, 91.
60. **Bertok, L.,** Radio-detoxified endotoxin as a potent stimulator of nonspecific resistance, *Perspect. Biol. Med.*, 24, 1980.
61. **Bertok, L.,** Stimulation of nonspecific resistance by radiation-detoxified endotoxin, in *Beneficial Effects of Endotoxins*, Nowotny, A., Ed., Plenum Press, New York, 1983, 213.
62. **Greisman, S. E.,** Induction of endotoxin tolerance, in *Beneficial Effects of Endotoxins*, Nowotny, A., Ed., Plenum Press, New York, 1983, 149.
63. **Braude, A. I.,** Absorption, distribution and elimination of endotoxins and their derivatives, in *Bacterial Endotoxins*, Landy, M. and Braun, W., Eds., Rutgers University Press, New Brunswick, NJ, 1964, 98.
64. **Greisman, S. E., Hornick, R. B., and Woodward, T. E.,** The role of endotoxin during typhoid fever and tularemia in man. III. Hyperreactivity to endotoxin during infection, *J. Clin. Invest.*, 43, 1747, 1964.
65. **Greisman, S. E., Hornick, R. B., Carozza, F. A., Jr., and Woodward, T. E.,** The role of endotoxin during typhoid fever and tularemia in man. I. Acquisition of tolerance to endotoxin, *J. Clin. Invest.*, 42, 1064, 1963.
66. **McCabe, W. R.,** Endotoxin tolerance. II. Its occurence in patients with pyelonephritis, *J. Clin. Invest.*, 42, 618, 1963.
67. **Rubenstein, M., Mullholland, J. H., Jeffrey, G. M., and Wolff, S. M.,** Malaria induced endotoxin tolerance, *Proc. Soc. Exp. Biol. Med.*, 118, 283, 1965.
68. **Abernathy, R. S. and Spink, W. W.,** Studies with Brucella endotoxin in humans: the significance of susceptibility to endotoxin in the pathogenesis of brucellosis, *J. Clin. Invest.*, 37, 219, 1958.
69. **Madonna, G. S., Peterson, J. E., Ribi, E. E., and Vogel, S. N.,** Early phase endotoxin tolerance: induction by a detoxified lipid A derivative, monophosphoryl lipid A, *Infect. Immun.*, 52, 6, 1986.
70. **Apte, R. N., Galanos, C., and Pluznik, D. H.,** Lipid A, the active part of bacterial endotoxin in inducing serum colony activity and proliferation of splenic granulocyte/macrophage progenitor cells, *J. Cell. Physiol.*, 87, 71, 1976.
71. **Metcalf, D.,** The molecular biology and functions of the granulocyte-macrophage colony-stimulating factors, *Blood*, 67, 257, 1986.
72. **Pluznik, D. H.,** Endotoxin-induced release of colony-stimulation factor, in *Beneficial Effects of Endotoxins*, Nowotny, A., Ed., Plenum Press, New York, 1983, 397.
73. **Bradley, T. R. and Metcalf, D.,** The growth of mouse bone marrow cells *in vitro*, *Aust. J. Exp. Biol. Med. Sci.*, 44, 287, 1966.
74. **Ho, M.,** Induction of interferon by endotoxin, in *Beneficial Effects of Endotoxins*, Nowotny, A., Ed., Plenum Press, New York, 1983, 381.
75. **Ho, M.,** Interferon like viral inhibitor in rabbits after intravenous administration of endotoxin, *Science*, 146, 1472, 1964.
76. **Stinebring, W. R. and Youngner, J. S.,** Patterns of interferon appearance in mice injected with bacteria or bacterial endotoxin, *Nature*, 204, 712, 1964.
77. **Sauter, C. and Wolfsberger, C.,** Interferon in human serum after injection of endotoxin, *Lancet*, 2, 852, 1980.
78. **Youngner, J. S. and Feingold, D. S.,** Interferon production in mice by cell wall mutants of *Salmonella typhimurium*, *J. Virol.*, 1, 1164, 1967.
79. **Blanchard, D. K., Djeu, J. Y., Klein, T. W., Friedman, H., and Stewart, W. E.,** II, Interferon-induction by lipopolysaccharide: dependence on interleukin 2 and macrophages, *J. Immunol.*, 136, 963, 1986.
80. **Michie, H. R., Manogue, K. R., Spriggs, D. R., Revhaugh, A., O'Dwyer, S., Dinarello, C. A., Cerami, A., Wolff, S. M., and Wilmore, D. W.,** Detection of circulating tumor necrosis factor after endotoxin administration, *N. Engl. J. Med.*, 318, 1481, 1988.
81. **Dinarello, C. A., Goldin, N. P., and Wolff, S. M.,** Demonstration and characterization of two distinct human leukocytic pyrogens, *J. Exp. Med.*, 139, 1369, 1974.

82. **Atkins, E. A. and Bodel, P.,** Fever, *N. Engl. J. Med.*, 286, 27, 1972.
83. **Waisbren, B. A.,** Bacteremia due to Gram-negative bacilli other than Salmonella. A clinical and therapeutic study, *Arch. Intern. Med.*, 88, 467, 1951.
84. **Yoshikawa, T., Tanaka, K. R., and Guze, L. B.,** Infection and disseminated intravascular coagulation, *Medicine*, 50, 237, 1971.
85. **Shands, J. W., Jr., Miller, V., and Martin, H.,** The hypoglycemic activity of endotoxin, I, occurrence in animals hyperreactive to endotoxin, *Proc. Soc. Exp. Biol. Med.*, 130, 413, 1969.
86. **Beutler, B., Mahoney, J., Trang, N. L., Pakala, P., and Cerami, A.,** Purification of cachectin, a lipoprotein lipase-suppressing hormone secreted by endotoxin-induced RAW 264.7 cells, *J. Exp. Med.*, 161, 984, 1985.
87. **Beutler, B., Greenwald, D., Holmes, J., Chang, D., Pan, T. C. E., et al.,** Identity of tumor necrosis factor and the macrophage-secreted factor cachectin, *Nature*, 316, 552, 1985.
88. **Cerami, A. and Beutler, B.,** The role of cachectin/TNF in endotoxin shock and cachexia, *Immunol. Today*, 9, 28, 1988.
89. **Urbaschek, B. and Forsmann, W.,** The endothelial cell of the microvascular bed as a target of endotoxic activities, in *The Reticuloendothelial System and the Pathogenesis of Liver Disease*, Liehr, H. and Grum, M., Eds., Elsevier, Amsterdam, 1981, 401.
90. **Hsuch, W., Gonzales-Crussi, F., and Arroyave, J. L.,** Platelet-activating factor-induced ischemic bowel necrosis. An investigation of secondary mediators in its pathogenesis, *Am. J. Pathol.*, 122, 231, 1986.
91. **Larsen, G. L. and Henson, P. M.,** Mediators of inflammation, *Annu. Rev. Immunol.*, 1, 335, 1983.
92. **Dinarello, C. A., Cannon, J. G., Mier, J. W., Bernheim, H. A., and LoPreste, G.,** Multiple biological activities of human recombinant interleukin 1, *J. Clin. Invest.*, 77, 1734, 1986.
93. **Bevilacque, M. P., Pober, J. S., Wheeler, M. E., Cotran, R. S., and Gimbrone, M. A.,** Interleukin-1 activation of vascular endothelium: effects on procoagulant activity and leukocyte adhesion, *Am. J. Pathol.*, 121, 394, 1985.
94. **Beutler, B., Krochin, M., Milsart, I. W., Luedke, C., and Cerami, A.,** Control of cachectin (tumor necrosis factor) synthesis: mechanisms of endotoxin resistance, *Science*, 232, 977, 1986.
95. **Gardner, K. D., Jr., Reed, W. P., Evan, A. P., Zedalis, J., Hylarides, M. D., and Leon, A. A.,** Endotoxin provocation of experimental renal cystic disease, *Kidney Int.*, 32, 329, 1987.
96. **Miller, M. A., Prior, R. B., Horvath, F., and Hjelle, J. T.,** Comparison of the *Limulus* Amoebocyte Lysate (LAL) assay and urine culture for detection of renal bacteriuria in polycystic kidney disease (PKD), in *Abstr. Ann. Meet. Am. Soc. Microbiol.*, American Society of Microbiology, Washington, D.C., 1988, 358.
97. **Miller, M. A., Prior, R. B., Horvath, F., and Hjelle, J. T.,** Detection of endotoxiuria in polycystic kidney disease patients by the use of the *Limulus* amoebocyte lysate assay, *Am. J. Kid. Dis.*, 15, 117, 1990.
98. **Proctor, R. A.,** Introduction, in *Handbook of Endotoxin*, Vol. 4, *Clinical Aspects of Endotoxin Shock*, Proctor, R. A., Ed., Elsevier, Amsterdam, 1986, xiii.
99. **Jorgensen, J. H.,** Clinical applications of the Limulus amebocyte lysate test, in *Handbook of Endotoxin*, Vol. 4, *Clinical Aspects of Endotoxin Shock*, Proctor, R. A., Ed., Elsevier, Amsterdam, 1986, 127.
100. **McCracken, G. H., Jr., Mustafa, M. M., Rami, O. O., Olsen, K. D., and Risser, R. C.,** Cerebrospinal fluid interleukin 1-beta and tumor necrosis factor concentration and outcome from neonatal gram-negative enteric bacillary meningitis, *Pediatr. Infect. Dis. J.*, 8, 155, 1989.
101. **Lebel, M. H., Freij, B. J., Syrogiannopoulos, G. A., Chrane, D. F., Hoyt, M. J., Stewart, S. M., Kennard, B. D., Olsen, K. D., and McCracken, G. H., Jr.,** Dexamathasone therapy for bacterial meningitis, *N. Engl. J. Med.*, 319, 964, 1988.

Chapter 2

THE AMERICAN HORSESHOE CRAB, *LIMULUS POLYPHEMUS**

Carl N. Shuster, Jr.

TABLE OF CONTENTS

I. Introduction 16

II. Body Sections and Features 16

III. Major Organ Systems 18

IV. Hemolymph 19

V. Long Life Cycle 21

VI. Environmental and Biological Interfaces 24

Selected References 25

* VIMS Contribution No. 1497.

I. INTRODUCTION

The American horseshoe crab, *Limulus polyphemus*, is, like its close relatives (three other species in the Indo-Pacific coastal area) a biological wonder. Its relatively simple yet complex biological characteristics make it better suited for a greater diversity of medical and physiological research than any other marine invertebrate. For example, experimentation on comparisons of the effects of different ions and drugs on nerve/heart muscle preparations is possible because the heart can be stripped of its nerves.

Complex biological systems enable *Limulus* to survive within wide ranges of environmental conditions. This is remarkable, particularly since outwardly, horseshoe crabs have apparently changed very little in the past 300 million years. Their present genetic makeup may be the result of retaining "ancestral genes" evolved throughout eons of environmental changes. Thus, by not discarding an evolved ability to conform to a wide range of conditions, the species reached a high plateau of generalization. Whatever has occurred, today *Limulus* is tolerant of wide ranges of temperature, salinity, substrate, diet, exposure to air, etc. This tolerance offers a clue to the lengthy, albeit intermittent, latitudinal distribution of the American horseshoe crab along the Atlantic coast, from Maine to Yucatan. There appears to be only one major flaw; they need wave-protected beaches bathed by tides of sufficient amplitude (as found within most estuaries) in which to lay their eggs. Furthermore, intertidal flats or extensive shoal water areas in the vicinity of spawning beaches are apparently essential for the development of large populations.

The biological characterization of any one population of horseshoe crabs has never been completed. Therefore, the following account, which concentrates on the natural history/ecology of *Limulus*, is a generalization. It is compiled from studies at different locations conducted at different times by several researchers. This is why such a generalization is likely to be challenged by anyone studying a specific population. For it is well known that there are morphometrically discrete populations along the coast and there may also be physiological subspecies. Thus, it is not surprising to find that each population is slightly different from others in its genetic makeup, as evidenced by its morphometry, biochemistry, physiology, or behavior.

II. BODY SECTIONS AND FEATURES

One's first impression, when seeing a horseshoe crab lumbering across a stretch of beach toward water, is that it is a miniature bulldozer. A closer look reveals that the foremost part appears to be shaped like a horse's foot. Indeed, colonial fishermen referred to this animal as a "horse foot crab" or the plural "horsefeet". "Horseshoe crab", the name most commonly used today, was probably derived from the horseshoe-like appearance of the under surface of the animal's front part. Other colorful names once used include king crab, sword tail crab, or sauce pan or pan crab (because the empty shell was used to help bail out rowboats). In fact, however, these animals are not crabs, i.e., crustacea. Their feeding appendages include chelicera and their respiratory appendages carry hundreds of lamellae; these are among several characteristics indicating a closer relationship to scorpions and spiders.

The streamlined body shape is suitable for burrowing and locomotion, although both activities are cumbersome. Horseshoe crabs swim gracefully, however, upside down at the surface of calm water. As they swim, their legs and flaplike appendages (operculum and respiratory) move in a "sculling" motion. The tripartite body is enclosed by a chitinous exoskeleton (shell or carapace). Some of the more prominent external features are described in the following text and illustrated in the accompanying diagrams. (Figure 1 describes the major features of the external surface of the body; Figure 3D [see also Figure 4A] shows the appendages which are aligned in sequence along the ventral axis of the body.)

The front part, the horsefoot-shaped prosoma, combines head and thoracic segments. Its most

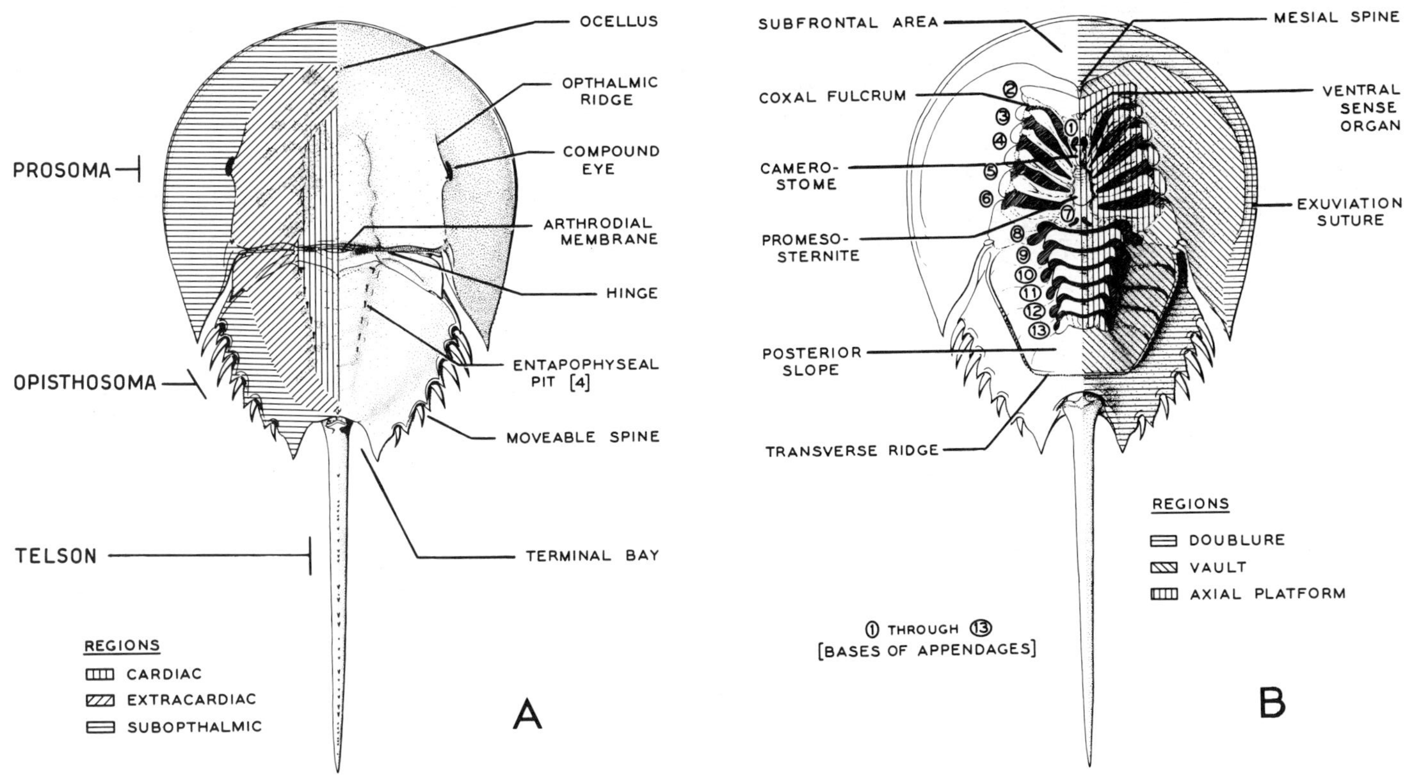

FIGURE 1. Adult male *Limulus polyphemus*. (A) Dorsal aspect, showing anatomical/topographical characteristics of the three major divisions of the body: prosoma, opisthosoma, and telson. The right half is stippled to show contour of the exoskeleton. The entapophyseal pits on the opisthosoma extend internally as projections to which the branchial muscles are attached. (B) Ventral aspect. The appendages (see Figures 3D and 4A) are omitted for clarity of the axial area. The bases of the appendages are identified by an encircled number placed just above or to the left of the bases (indicated by heavy cross-hatching). The right side is stippled to suggest exoskeleton contours. (Modified from Shuster, C.N., Jr., in *Physiology and Biology of Horseshoe Crabs,* Bonaventura, J., et al., Eds., Alan R. Liss, New York, 1982, 1.)

prominent dorsal features, besides a number of spiny projections, are a pair of large, lateral compound eyes and a pair of simple eyes, located close together, up front. The spines are longer and sharper in the juveniles, blunter in larger animals and adults. Underneath, along the axis of the body, are seven pairs of appendages. The first pair, which are small and most anteriorward (chelicera: Figures 1B and 3D, appendage 1) and the seventh pair, small axially situated "paddles" (chilaria), function mainly in feeding, while the larger appendages in between are used in both feeding and locomotion. The pincers on the first four pairs of walking legs (Figure 1B, located at 2 to 5; see also Figure 3D) help guide food to the centrally located mouth. The chelicera push food backwards, the chilaria, forward. The last pair of legs (Figure 3D, 6) are specialized, with blades on the last segment that expand when pressed against the substrate, helping to propel the animal forward. The last prosomal appendage is a large flap, the operculum (Figure 3D, 8), which covers the respiratory appendages (borne on the next portion of the body). The operculum bears a pair of openings from the reproductive organs near its posterior base: hard, conical papillae on the male, soft, oval openings on the female.

The middle section, the opisthosoma, is the main abdominal part. Its large, ventral vault is filled with the five respiratory appendages (Figure 1B, at 9 to 13). Each of the appendages, actually fused pairs, bear two sets of lamellae or "books". Each book, in the larger animals, has over 100 oval-shaped, very thin lamellae in "stacks" of decreasing size on its posterior face. Gaseous exchanges occur through these lamellae, commonly referred to as "gills"; hence the term "book gills".

The last part, the telson, is sword-like. Its origin may have been in the articulation of an ancestral posterior, nonmoveable axial spine. It swivels throughout a 360° conical area and is used in locomotion, especially to thrust off the bottom to gain a swimming posture at the surface. Individuals stranded on beaches and tidal flats at ebb tide are often able to right themselves, using the telson as a lever and its base as a fulcrum, by kicking their legs while "arching" the prosoma and opithosoma dorsally, at the hinge. The telson is not a "stinger" or weapon as might at first be imagined. It does, however, provide the easiest means to pick up a horseshoe crab.

III. MAJOR ORGAN SYSTEMS

Internally, the organ systems are much intermingled as they ramify from their central, axial positions throughout the rest of the body cavity. Their relative positions are most distinct along the axis. The heart is in the most dorsal position, just beneath the carapace (Figure 1A, the cardiac region); next is the intestine and the reproductive ducts which parallel it, while the central nervous system is in the most ventral position. Comparison of the axial position of the arthrodial membrane in the hinge area (Figure 1A) with the location of the heart (Figure 2) illustrates why this area is the best access point for removing blood with a syringe. The tubular heart lies within a sinus into which the blood flows, after collecting in large venous sinuses, from the respiratory lamellae. The blood then passes into the heart through several pairs of ostia and is pumped laterally and forward through arteries.

The digestive tract has two distinct parts, a tubular portion and a large mass of digestive diverticulae. Food passes anteriorally from the centrally located mouth (Figure 3D, at the bases of appendages 2 and 3) through the chitinous-lined esophagus into the proventriculus ("crop" or "gizzard"). The latter has knobby, chitinous ridges which aid in trituration of food masses. The resulting material, sometimes even whole, small mollusks (up to 7 mm in length), is then passed through a pyloric value into the tubular midgut. Two pairs of openings lead from the forward portion of the midgut into multibranched digestive diverticulae ("liver" masses) where intracellular digestion occurs. The "liver" tissue is intermingled with the reproductive structures. Fecal material is encased in mucous prior to discharge via a chitinous-lined rectum.

The reproductive system of the mature female is comprised of multibranching, thin-walled tubules of decreasing diameters, filled with ripe eggs. Diffuse ovarian tissue, mostly in the

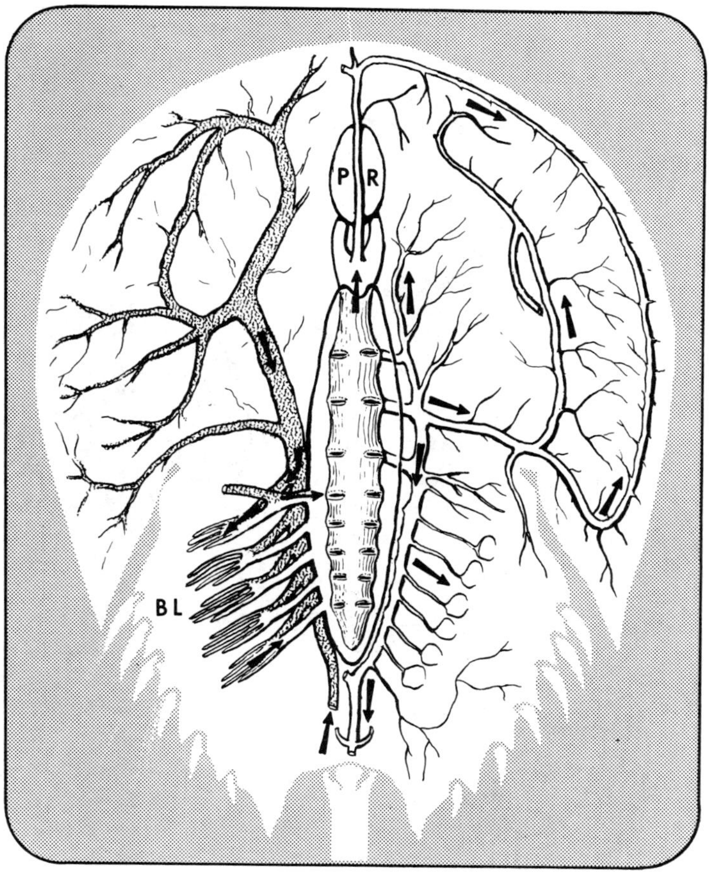

FIGURE 2. A diagram of the circulatory/respiratory system. The outline of the body is shaded to provide a frame of reference. While the circulatory system is symmetrical, only the axially located heart, within its sinus, is completely shown; the venous system of collecting tubules is shown on the left, arteries on the right. Eight pairs of large ostia open and close synchronously with an anteriorward peristaltic pulsation of the heart, moving blood from the sinus into the striated heart. Blood passes anteriorly from the heart into three arteries, two of which pass downward behind the proventriculus (PR) and form a ring enclosing the brain. Three pairs of main, lateral arteries carry blood throughout the rest of the body. Blood returns to the sinus surrounding the heart via the five "books" of respiratory lamellae (BL). For simplicity, only the roots of the lesser arteries and portions of the veins are shown. Neither the arteries to the brain (beneath the proventriculus) nor the ventral artery are shown. Arrows indicate direction of blood flow.

peripheral area of the body cavity, is evident by masses of almost microscopic eggs. The gonoducts of the male are less visible; mature males carry sperm throughout the year.

The entire central nervous system is encased within arteries. A "cartilaginous", shield-shaped, protective plate lies above the doughnut-shaped "brain".

IV. HEMOLYMPH

The hemolymph of *Limulus* contains hemocyanin, a copper-based respiratory protein with two copper atoms per oxygen binding site, dissolved in the lymph and one or two types of blood cells (depending upon the authority consulted). These cells are motile, hence amoebocytes, and are as much as 4%, by volume, of the hemolymph. They vary in shape from ellipsoid to round,

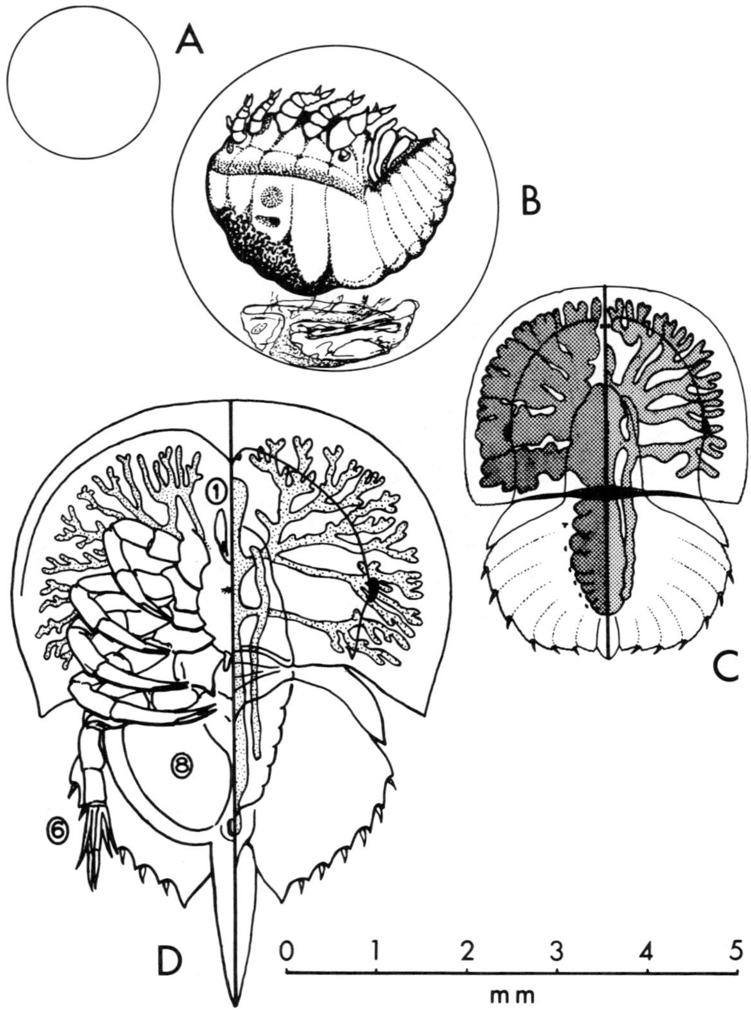

FIGURE 3. Early stages in the life cycle of *Limulus polyphemus*. (A) The water-hardened, newly laid egg. (B) The developing embyro in its transparent capsule. Note the exuviae beneath the embryo. The first ten pairs of appendages are already in evidence; the prosomal appendages appear first, followed by the branchiae. The yolk is depicted by heavy stippling (lower left quadrant). The round organ below (above in the drawing because the embryo is upside down), the elliptical eye, disappears by the first-tailed stage (D). (C) Early and late development of the digestive tract in the larval stage. At first the digestive tract is still embryonic. The left side shows the segmental arrangement of the largely undigested yolk; the right side shows a nearly complete system (compare it with the next stage, in D). (D) The first-tailed stage. The left side shows the ventral view, with the walking legs pulled away from the axial region to show the central position of the mouth (opposite the bases of the second and third pair of appendages). The anus is at the base of the telson. The dorsal view (right) shows the extent of the digestive system (anterior-most, note the bulbous appearance of the dorsal aspect of the proventriculus). From the sparsely branched condition of the diverticulae in these early stages (C and D), the multibranched diverticulae continue development, from molt to molt, resulting in the massive structures comprised of tiny tubules seen in the larger animals. The positions of an ocellus and one of the elliptical, compound eyes, are shown on the opthalmic ridge, in their respective positions, axially (at front) and laterally. Appendages are numbered as in Figure 1B: 1 = chelicera, 6 = the fifth, "pusher" leg, and 8 = the operculum.

and are nucleated and granular. The amoebocytes circulate throughout the body cavity, a phenomenon which can be demonstrated *in vivo* by microscopic observation through the thin chitinous carapace of the first-tailed stage (Figure 3D).

The amoebocytes have several functions. They act as phagocytes and perform intracellular digestion and store and transport digested materials. They serve as a defense mechanism in the repair of wounds and in clotting. The blood clotting system is based solely on a sequential degradational response of the amoebocytes to foreign material, such as an endotoxin; the sequence consists of rapid aggregation of the cells, followed by pseudopod projection in all directions, and finally release of their cytoplasm.

Until withdrawn from the animal, the hemolymph is a pale, slightly straw colored liquid. Upon exposure to air, it turns a dark blue. Hemocyanin is the only protein in the serum of the horseshoe crab after complete removal of the clotted amoebocytes. Clotting does not remove an appreciable amount of hemocyanin from the serum. Variation in hemocyanin and amoebocyte percent is due to fluctuations in the water content of the serum during the molting process and during changes in environmental salinity. The hemolymph is isomotic or slightly hypotonic at the higher salinities and hyperosmotic in low salinities.

V. LONG LIFE CYCLE

Spawning, the most spectacular event in the life cycle of horseshoe crabs, occurs every year. Responding to seasonal stimuli, the adults migrate from the deeper waters where they spend most of the year to protected beaches within estuaries. Depending in part upon latitude, they arrive in peak numbers on the beaches at the time of the full moon tides, in late spring to early summer. In many places isolated mating groups occur along a beach, but in others, such as Delaware Bay, the beach is literally paved with them (Figure 4B).

Generally, there are more males in the mating groups. They tend to "patrol" along the foot of the beaches, awaiting the arrival of the females which go directly to the beach. When the females pass through this "stag line", the males sense them and move further onto the beach. Only adult males have a special pair of clasping appendages that enables them to attach to the posterior projections of the female opisthosoma on either side of the telson. Only one male occupies the central, attached position; excess males mill closely about. During mating the males are continuously milting and fertilization occurs externally.

A large female may contain over 88,000 eggs, which she lays in several nests in holes about 15 cm deep, that she digs into the beach. The extruded eggs are small, soft, sticky, and usually greenish in color. In contact with seawater they harden into tiny opaque spheres (Figure 3A). When she has completed laying her eggs over one or more days during the high tides, the female generally leaves the area. However, some stay in the shallow water, especially if food is abundant. Males may stay attached to their spawning partner for some time while the unattached "bachelors" may return to their patrol along the foot of the beach. Such behavior, especially when large numbers of animals are involved, is one reason why males outnumber females on the beaches.

Within the nests, embryonic development of the fertilized eggs is rapid and may be completed within 2 weeks. For the first few days the egg covering is opaque. Then the original covering splits and is sloughed off and a new, transparent covering encloses the embryo (Figure 3B). The embryos shed their thin, embryonic shells several times. Activity of the embryonic appendages begins as soon as the joints are formed and continues incessantly thereafter. The heavier yolk keeps the center of gravity low and the appendages uppermost. Thus, the larvae have a lot of "practice" in upside-down swimming before they escape from the nests. They are washed out of the nest but they participate in their liberation. Sensing the rising tide, the whole nest of larvae, acting in a concerted effort, dig their way to the surface of the beach. Once waterborne they swim upwards and then drift down, repeating this until the time of molting, in about 6 d, when they

FIGURE 4. (A) An immature horseshoe crab, about 12 cm in width, in the early stage of emerging from old carapace. It was removed from an aquarium and turned over to photograph the molting process. Note the new animal emerging forward from a slit in the rim of the exoskeleton. At this stage the animal has the consistency of a firm foam rubber.

settle on the bottom. Thereafter they are mainly benthic although they retain their swimming ability throughout life.

At the time of hatching, embryonic development, particularly of the digestive tract, is incomplete. As soon as the digestive tract is fully formed, the tiny larvae are ready to molt (Figure 3C). This next stage is the "first-tailed" instar; they resemble miniature adults except that the telson is the axial, terminal spine of the body (Figure 3D). The telson becomes articulated at the next molt. Molting is frequent at first, with the soft animal "extruding" itself forward through a horseshoe-shaped slit that opens around the ventral rim (Figure 4A). The frequency of molting is a function of age. It occurs during the summer, up to four or five times the first year, then two or three times the next, a couple of times the third year, and thereafter, once a year. The smaller, hence the younger, the animals, the more rapid the emergence from the old shell. Exuviation takes only a few minutes for the larvae and several hours for nearly full-grown animals. Although soft and pliable, the newly molted animals are capable of moving about.

At first the young horseshoe crabs remain in shallow water near the beaches from which they hatched. As they grow larger, they move further away, until as adults, especially if "pressured" by large numbers of their peers, they move out from the estuaries onto the continental shelf. At present it is believed that males mature in about 15 stages, females in 16, at about 9 and 10 years of life, respectively. That would account for the larger average size of adult females, who are 25 to 30% larger in prosomal width than the average males, unless the females also grow slightly larger at each instar. Many adults live at least 5 years, and some may reach a total age of 20 years.

Only the males show secondary sexual characteristics, which appear at their last molt. The chief changes are in the first pair of walking legs and the anterior portion of the prosoma, which becomes arched, enabling the male prosoma to override the female opisthosoma. When the pincers of the first leg are modified, they change into a bulbous part against which the moveable chela (now greatly curved) closes. The nonmoveable chela atrophies and breaks off. The result is the "fist-and-thumb" clasping organ used to achieve amplexus in mating.

20. **Okuguchi, S.,** Improvement of the micromethod for the Limulus lysate test, *Microbiol. Immunol.*, 22, 113, 1978.
21. **Hollander, V. P. and Harding, W. C.,** A sensitive spectrophotometric method for measurement of plasma endotoxin, *Biochem. Med.*, 15, 28, 1976.
22. **Nandan, R. and Brown, D. R.,** An improved in vitro pyrogen test: to detect picograms of endotoxin contamination in intravenous fluids using Limulus amoebocyte lysate, *J. Lab. Clin. Med.*, 89, 910, 1977.
23. **Munford, R. S.,** Quantitative *Limulus* lysate assay for endotoxin activity: aggregation of radioiodinated coagulation monomers, *Ann. Biochem.*, 91, 509, 1978.
24. **Goto, H. and Nakamura, S.,** Dry up method as a revised Limulus test with a new technique for gelation inhibitor removing, *Jpn. J. Exp. Med.*, 49, 19, 1979.
25. **May, S. A., Mikolajcik, E. M., and Richter, E. R.,** Conventional tube and microplate *Limulus* amoebocyte lysate procedures for determination of gram-negative bacteria in milk, *J. Dairy Sci.*, 72, 1137, 1989.
26. **Prior, R. B. and Spagna, V. A.,** Adaptation of a microdilution procedure to the *Limulus* lysate assay for endotoxin, *J. Clin. Microbiol.*, 10, 394, 1979.
27. **Tsuji, K. and Harrison, S. J.,** Limulus amebocyte lysate — a means to monitor inactivation of lipopolysaccharide, in *Biomedical Applications of the Horseshoe Crab (Limulidae)*, Cohen, E., Ed., Alan R. Liss, New York, 1979, 367.
28. **Jorgensen, J. H. and Alexander, G. A.,** Automation of the *Limulus* amoebocyte lysate test by using the Abbott MS-2 microbiology system, *Appl. Environ. Microbiol.*, 41, 1316, 1981.
29. **Albaugh, B. R. and Chandler, C. B.,** Automated methodology for the Limulus amebocyte lysate (LAL) assay using the multiskan microplate reader, in *Endotoxins and Their Detection with the Limulus Amebocyte Lysate Test*, Watson, S. W., Levin, J., and Novitsky, T. J., Eds., Alan R. Liss, New York, 1982, 183.
30. **Novitsky, T. J., Remillard, J. F., and Loy, N.,** Design criteria and evaluation of the LAL-4000 for kinetic turbidimetric LAL assay, in *Detection of Bacterial Endotoxins with the Limulus Amebocyte Lysate Test*, Watson, S. W., Levin, J., and Novitsky, T. J., Eds., Alan R. Liss, New York, 1987, 189.
31. **Novitsky, T. J., Roslansky, P. F., Siber, G. R., and Warren H. S.,** Turbidimetric method for quantifying serum inhibition of *Limulus* amoebocyte lysate, *J. Clin. Microbiol.*, 20, 211, 1985.
32. **Iwanaga, S., Morita, T., Harada, T., Nakamura, S., Niwa, M., Takada, T., Kimura, T., and Sakakibara, S.,** Chromogenic substrates for horseshoe crab clotting enzyme — its applications for the assay of bacterial endotoxins, *Haemostasis*, 7, 183, 1978.
33. **Harada, T., Morita, T., Iwanaga, S., Nakamura, S., and Niwa, M.,** A new chromogenic substrate method for assay of bacterial endotoxins using *Limulus* hemocyte lysate, in *Biomedical Applications of the Horseshoe Crab (Limulidae)*, Cohen, E., Ed., Alan R. Liss, New York, 1979, 209.
34. **Roth, R. I., Levin, J., and Behr, S.,** A modified *Limulus* amebocyte lysate test with increased sensitivity for detection of bacterial endotoxin, *J. Lab. Clin. Med.*, 114, 306, 1989.
35. **Ikemura, K., Ikegami, K., Shimazu, T., Yoshioka, T., and Sugimoto, T.,** False-positive result in *Limulus* test caused by *Limulus* amebocyte lysate-reactive material in immunoglobulin products, *J. Clin. Microbiol.*, 27, 1965, 1989.
36. **Morita, T., Tanaka, S., Nakamura, T., and Iwanaga, S.,** A new (1–3)-β-D-glucan-mediated coagulation pathway found in limulus amebocytes, *FEBS Lett.*, 129, 318, 1981.
37. **Friberger, P., Sorskog, L., Nilsson, K., and Knos, M.,** The use of a quantitative assay in endotoxin testing, in *Endotoxins and Their Detection with the Limulus Amebocyte Lysate Test*, Watson, S. W., Levin, J., and Novitsky, T. J., Eds., Alan R. Liss, New York, 1987, 149.
38. **Levin, J.,** The history of the development of the *Limulus* amebocyte lysate test, in *Bacterial Endotoxins, Structure, Biochemical Significance, and Detection with the Limulus Amebocyte Lysate Test*, Cate, J. W., Buller, H. R., Sturk, A., and Levin, J., Eds., Alan R. Liss, New York, 1985, 3.
39. **Obayashi, T., Tamura, H., Tanaka, S., Ohki, M., Takahashi, S., Arai, M., Masuda, M., and Kawai, T.,** A new chromogenic endotoxin-specific assay using recombined *Limulus* coagulation enzymes and its clinical applications, *Clin. Chim. Acta*, 149, 55, 1985.
40. **Prior, R. B. and Finn, J.,** unpublished data, 1988.
41. **Lindsay, G. K., Roslansky, P. F., and Novitsky, T. J.,** Single-step, chromogenic *Limulus* amebocyte lysate assay for endotoxin, *J. Clin. Microbiol.*, 27, 947, 1989.
42. **Jorgensen, J. H. and Alexander, G. A.,** Automation of the *Limulus* amoebocyte lysate test by using the Abbott MS-2 microbiology system, *Appl. Environ. Microbiol.*, 41, 1316, 1981.
43. **Oishi, H., Takaoka, A., Hatayama, Y., Matsuo, T., and Sakata, Y.,** Automated Limulus Amebocyte Lysate (LAL) Test for Endotoxin Analysis using a New Toxinometer ET-201, *J. Parent. Sci. Technol.*, 39, 194, 1985.
44. **Baek, L.,** New, sensitive rocket immunoelectrophoretic assay for measurement of the reaction between endotoxin and *Limulus* amoebocyte lysate, *J. Clin. Microbiol.*, 17, 1013, 1983.
45. **Baek, L., Holby, N., Hertz, J. B., and Espersen, F.,** Interaction between *Limulus* amoebocyte lysate and soluble antigens from *Pseudomonas aeruginosa* and *Staphylococcus aureus* studied by quantitative immunoelectrophoresis, *J. Clin. Microbiol.*, 22, 229, 1985.

46. **Zhang, G.-H., Baek, L., and Koch, C.,** New microassay for quantitation of endotoxin using *Limulus* amebocyte lysate combined with enzyme-linked immunosorbent assay, *J. Clin. Microbiol.*, 26, 1464, 1988.
47. **Kakinuma, A., Asano, T., Torii, H., and Sugino, Y.,** Gelation of *Limulus* amebocyte lysate by an antitumor (1–3)-β-D-glucan, *Biochem. Biophys. Res. Commun.*, 101, 434, 1981.

Chapter 4

ROLE OF THE FDA IN REGULATING THE *LIMULUS* AMOEBOCYTE LYSATE TEST

H. Donald Hochstein

TABLE OF CONTENTS

I. Introduction 38
   A. Background of the LAL Testing in FDA 38
   B. First Uses of the LAL Test 39
   C. Licensing of the LAL Test 39
   D. FDA's Experience with Endotoxin 40
   E. Federal Register Notices Leading to the LAL Guideline 43
   F. Legal Effect of the Guideline 45
   G. LAL Licensed Manufacturers 45

II. Methods Approved by FDA for the LAL Test 46
   A. Gel End Point Method 46
   B. Chromogenic Substrate Method 46
   C. Turbidimetric Method 46

III. Automated Systems Approved by FDA 46
   A. The LAL-5000 and the Kinetic Turbidimetric Assay 46
   B. The Toxinometer ET-201 47

IV. Labeling the LAL for Testing Body Fluids for Endotoxin 47

V. Testing U.S. Standard Endotoxin (Lot EC-5) in Rabbits and Humans 48

VI. Summary 49

References 49

I. INTRODUCTION

The U.S. Food and Drug Administration (FDA) is responsible for licensing all *Limulus* amoebocyte lysate (LAL) manufacturers. In order to be licensed, a manufacturer must complete **Form FDA 3210** (Application for Establishment License for Manufacture of Biological Products) and **Form FDA 2567** (transmittal of labels and circulars). The manufacturer must also prepare three lots of lysate. Each lot must be tested for potency using the U.S. Standard Endotoxin, EC-5, which is identical to the U.S. Pharmacopeia (USP) Endotoxin Standard Lot F. The three lots must then be submitted to the Center for Biologics Evaluation and Research (CBER) for FDA testing. Once the testing is complete and the licensing committee has reviewed the application in detail, a prelicensing inspection of the actual manufacturing site or sites is scheduled. The licensing committee then reviews the results of both the inspection and the lot testing and a decision is made on granting an Establishment and Product License. Once a license is granted, the manufacturer must submit each lot of lysate prepared to CBER for testing and release. If all tests are passed, the manufacturer is notified that the lot is released and may be marketed.

A. BACKGROUND OF THE LAL TESTING IN FDA

In late 1970 a graduate student from Johns Hopkins University joined our laboratory at what was then the Division of Biologic Standards (DBS), National Institutes of Health (NIH) (later changed to the Bureau of Biologics [BOB], FDA.) This student was working on a masters degree under the direction of Dr. Jack Levin. His research required working with the lysate from the amoebocytes of the *Limulus polyphemus* (horseshoe crab). This material was said to be capable of detecting nanogram quantities of endotoxin (lipopolysaccharides (LPS) from the cell wall of Gram-negative bacteria). Since we were performing rabbit pyrogen testing on at least four licensed biological products per day, we decided to incorporate the LAL test when testing certain blood products. These products were 5 and 25% normal serum albumin (human), purified protein fraction (PPF), immune serum globulin (ISG), and antihemophilic factor (AHF).[1,2]

The first several batches of lysates were prepared at Johns Hopkins University. After about 6 months of trial and error, it was determined that the *in vitro* test was of value for the detection of endotoxin. Since we were about 150 mi from the ocean, amebocytes were difficult to obtain, so we decided to spend 1 week each year (for several years) at the NASA base in Chincoteague, VA bleeding horseshoe crabs. We would bleed 800 to 1000 crabs per week and return to the laboratory with several liters of lysate and 200 to 300 live crabs. These crabs were placed in our three-tier fiberglass lobster tanks, purchased from Bergson Products, Inc., of Warren RI (Figure 1). (Inclusion of manufacturers name does not imply endorsement by the Public Health Service of the U.S. Department of Health, Education and Welfare.) Each tank was 10 ft long and 20 in. deep. The tanks from bottom to top were 6, 5, and 4 ft wide. This provided total visibility of the interior of the bottom two tanks and also provided steps for reaching the top tank.

Each tank contained 7 in. of synthetic seawater (Instant Ocean). There was a pump connected to the bottom tank that pumped the salt water to the top tank. From there it ran out the overflow into the middle tank and then back to the bottom tank. Each tank held between 75 to 100 crabs depending on their size. If the tanks were full and we had extra crabs, we would put them in plastic boxes and place them in the refrigerator at 4 to 8°C. If the crabs were kept moist they would live in the refrigerator for at least 6 months without food or water. The crabs would not move in this environment as long as the light was off but when the light in the refrigerator was turned on the crabs would respond to this stimulus by moving their tails.

The crab tanks were housed in the rabbit holding room. Following a rabbit pyrogen test, the rabbits were sacrificed and their livers removed. These livers were rinsed in tap water, sliced, and fed to the crabs. The crabs seemed to thrive on this diet.

Each crab in the tanks was numbered so that we could keep track of the bleeding record. The

Reinforced Fiberglass Lobster Tanks

BPI Reinforced Fiberglass Lobster Tank

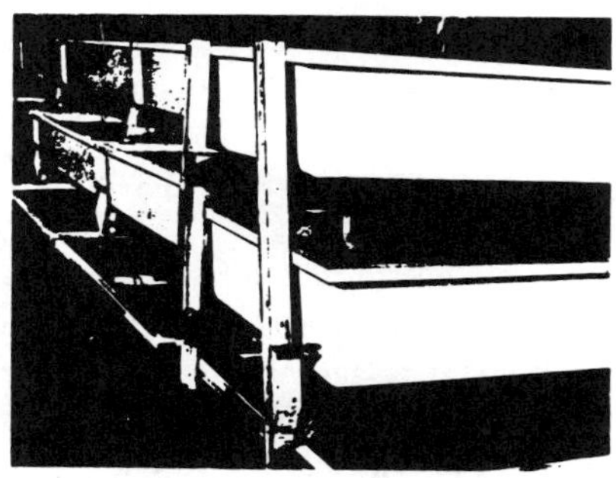

Note arrangement of tank sets on BPI stanchions.

FIGURE 1. Top shows a technician catching a horseshoe crab for bleeding. Bottom shows how the tanks are arranged to form steps.

crabs were bled on a semiannual basis. It was noted that when they were bled more frequently than every 6 months, the crabs produced hemolymph but very few amoebocytes, indicating that it took about 6 months for the horseshoe crab (in captivity) to regenerate its normal amoebocyte count.

B. FIRST USES OF THE LAL TEST

It was in 1972 that the LAL test began to be widely used in the medical community, and the FDA received complaints that some people were using lysate to diagnose septicemia, meningitis, and urinary tract infections. The FDA decided that regulations had to be written to control the sale and use of this product. Since LAL was a blood product and since the expertise for this product was in our Division (Division of Control Activities), it was decided to assign this product to the BOB.

C. LICENSING OF THE LAL TEST

In a **Federal Register Notice** of January 12, 1973 (38 FR 1404),[3] the FDA announced that LAL, derived from the circulating blood cells (amoebocytes) of the horseshoe crab (*Limulus polyphemus*), is a biological product. As such, it is subject to licensing requirements as provided in section 351 of the Public Health Service Act (42 U.S.C. 262). This notice also stated that all drugs subject to a pyrogen test are required to be tested by a rabbit pyrogen test and that the LAL test cannot be used as a final pyrogen test.

This notice prompted two manufacturers to apply, in 1976, for a license to produce LAL.

They completed the FDA Forms 3210 and 2567 for Establishment License for the Manufacture of Biological Products. These forms were submitted to the BOB along with samples from three final container lots complete with labeling. The following is a brief description of the licensing procedure.

When all of the forms and samples are received, an ad hoc committee is established to review the submission in detail. While the paper work is being reviewed, the samples are distributed to specific laboratories for moisture, sensitivity, and sterility testing. Once the form review and sample testing are complete, an on-site inspection of the manufacturing facility is scheduled. The prelicensing inspection is scheduled so that the appropriate personnel will be present for answering questions during the inspection. The inspector is usually one of the committee members with expertise in the testing of the product. The inspection involves not only checking the physical plant, but also reviewing daily records, including standard operating procedures for each production step. The validation records of equipment such as freeze-dryers, air flow hoods, and autoclaves are also reviewed. If any exceptions to the Federal Regulations are noted during the inspection, the inspector discusses them with management before leaving the facility. The manufacturer then has the responsibility to reply in writing to each observation. The licensing committee reviews all related data and submits a recommendation to the Director, CBER, regarding licensure.

Following the satisfactory in-house testing and prelicensing inspection, both manufacturers were licensed in 1977.

Once a manufacturer is licensed, each lysate lot submitted to the BOB for release must be accompanied by two forms, one of which is a protocol (Figure 2) that lists all the manufacturer's testing results, from sterility to chemical and physical data. The second form is the raw data potency sheet (Figure 3), which shows the test results on the contents of 20 final containers in parallel with 20 replicates of the FDA Reference Lysate. Both lysates are tested with U.S. Reference Endotoxin, Lot EC-5. Since we know the sensitivity of our Reference Lysate and the potency of Lot EC-5, we are able to tell if the testing laboratory is in control while determining the label sensitivity of the lysate.[4] Both the reference lysate and the endotoxin are supplied to LAL manufacturers for their in-house testing.

We test each lot of lysate for sensitivity before it is released. The difference between our testing and that of the manufacturer is that we use only four vials of lysate. If our test value is within twofold of the manufacturer's sensitivity test, the lot is released. If we are beyond the twofold range, we will test the contents of four more vials and average the results with the previous values. If our results are still outside these limits, we will continue testing until we have tested the contents of 24 vials.[4,5] If we are unable to confirm the manufacturer's test results at this level of testing, the lot is rejected and the manufacturer is notified that the lot cannot be released. The testing of samples and the review of the product protocol usually takes 2 to 3 weeks. We test every lot of lysate for confirmation of the labeled claim, while the sterility and moisture testing are usually conducted on a random basis.

D. FDA'S EXPERIENCE WITH ENDOTOXIN

It did not take long to realize that endotoxins prepared from different genera of bacteria, and even different lots prepared from the same genus gave different potencies with the same lysate on a per weight basis. Endotoxin lots varied in potency as much as 100 to 1000 times. This meant that the sensitivity of each LAL lot had to be prefaced with the endotoxin and lysate lot used so that another laboratory could repeat the results.[3]

The Bureau determined that, to control lots of lysate, both a standard endotoxin and a standard lysate were needed.[6] This prompted the Bureau to purchase a large lot of LAL from a licensed manufacturer and standardize it as U.S. Reference Lysate, Lot 1. The endotoxin was received from Dr. D. B. Castor of the Bureau of Drugs.[7] He had just completed a collaborative rabbit pyrogen study using *Klebsiella pneumoniae*. He supplied us with a small amount of this

PROTOCOL FOR LIMULUS AMEBOCYTE LYSATE

Manufacturer: License No. _______

Date of Protocol __________

Name of Product:

A) Proper Name: Limulus Amebocyte Lysate Lot No. _______

B) Trade Name:

C) Product Code:

Control Tests

A) Limulus amebocyte lysate

1) Sterility Direct XXX

a) Bulk Container: Date of Test: __________

Volume Tested: _________ Results: _____

b) Final Container: Date of Test: _________

Number of Units Tested: _____

Volume/Container Tested: ____ Results: __________

2) Potency Assay (Sensitivity)

Date on Test: _______ Test Volumes: 0.1 ml of endotoxin added to 0.1 ml of lysate with incubation at 37°C for 60 minutes.

FDA Reference Endotoxin: Lot No. EC-5

Endpoint Gelation Geometric Mean: __________ EU/ml
(raw data attached)

3) Chemical and Physical Data

a) Data Lyophilized: __________

b) Appearance Lyophilized:

c) Appearance Reconstituted:

d) pH: _______________

e) Residual Moisture: _________% Date on Test: _________

This is to certify that these are the results of all tests performed on this lot. These results indicate that the lot complies with all applicable requirements.

ALL THE ABOVE TESTS HAVE PASSED. ____________________
Responsible Head

FIGURE 2. Copy of the first page of the protocol that each lysate manufacturer must submit with each lot of *Limulus* lysate for release.

preparation, which we freeze-dried using 0.1% normal serum albumin (human) as a stabilizer. We labeled this as Lot 1b and it proved to be stable for several years.

In 1974, we decided to replace Lot 1b with an *Escherichia coli* endotoxin. A contract was awarded to Dr. J. A. Rudbach at the University of Montana. He prepared 30 g of endotoxin from *E. coli* 0113 (EC), using the hot aqueous phenol extraction method. The details of this EC preparation are outlined in a publication by Rudbach et al.[8] This bulk EC endotoxin is stored in a desiccator at room temperature and should meet our needs for many years if used only as a standard preparation.

In 1976, we used a portion of our bulk EC endotoxin to prepare Lot EC-1. Before freeze-drying, normal serum albumin (human) at 0.1% concentration was added as a stabilizer. This was a small lot consisting of 200 vials which lasted only about 1 year.[9]

We then decided to prepare a large lot of endotoxin in the same way we prepared EC-1. This

RAW DATA REPORT - LAL

Reference Lysate Lot 13
Endotoxin Lot EC-5
Concentrations are in EU/ml

Test Lysate Lot F27
Endotoxin Lot EC-5
Concentrations are in EU/ml

Test	0.5	0.25	0.125	0.06	0.03	E.Pt.	Vial	0.5	0.25	0.125	0.06	0.03	E.Pt.
1	+	+	+	-	-	.125	1	+	-	-	-	-	.5
1	+	+	+	-	-	.125	2	+	-	-	-	-	.5
1	+	+	+	+	-	.06	3	+	-	-	-	-	.5
1	+	+	+	-	-	.125	4	+	-	-	-	-	.5
1	+	+	+	-	-	.125	5	+	-	-	-	-	.5
1	+	+	+	+	-	.06	6	+	-	-	-	-	.5
1	+	+	+	+	-	.06	7	+	+	-	-	-	.25
1	+	+	+	-	-	.125	8	+	-	-	-	-	.5
1	+	+	+	-	-	.125	9	+	-	-	-	-	.5
1	+	+	+	+	-	.06	10	+	-	-	-	-	.5
2	+	+	+	-	-	.125	11	+	-	-	-	-	.5
2	+	+	+	-	-	.125	12	+	-	-	-	-	.5
2	+	+	+	-	-	.125	13	+	+	-	-	-	.25
2	+	+	+	+	-	.06	14	+	-	-	-	-	.5
2	+	+	+	-	-	.125	15	+	-	-	-	-	.5
2	+	+	+	-	-	.125	16	+	-	-	-	-	.5
2	+	+	+	-	-	.125	17	+	-	-	-	-	.5
2	+	+	+	-	-	.125	18	+	-	-	-	-	.5
2	+	+	+	-	-	.125	19	+	-	-	-	-	.5
2	+	+	+	-	-	.125	20	+	-	-	-	-	.5

MEAN (Log_2)- -3.265

S.D. (Log_2 Data)- .47

GEOMETRIC MEAN- .104

STATED END-POINT- .9 EU/ml

MEAN (Log_2)- -1.1

S.D. (Log_2 Data)- .31

GEOMETRIC MEAN- .46

STATED END-POINT- .5

FIGURE 3. Page two of a protocol that each lysate manufacturer must submit with each lot of *Limulus* lysate for release. The lysate reference Lot 13 is compared with the lot submitted for release.

lot we designated as EC-2 and 1500 vials were freeze-dried. Each vial contained 1.0 μg of EC endotoxin. After accumulating extensive data on this very stable standard preparation, we decided to assign a unit value to the preparation.[10] Most control authorities agree that it is better to use units of activity rather than a concentration based on dry weight in expressing the strength of reference preparations.[11] Based on a collaborative study involving BOB and several licensed LAL manufacturers, we assigned a value of 5.0 endotoxin units per nanogram (EU/ng) to Lot EC-2.[2]

The only criticism of Lot EC-2 was that it contained normal serum albumin (human) which might bind endotoxin in unknown products. In order to eliminate this hypothetical problem, in 1980 Lot EC-3 was prepared as a pilot lot with no fillers or stabilizers. Following several months of use, this lot appeared satisfactory and therefore Lot EC-4 was prepared in the same way. After freeze-drying, the potency of EC-4 was just half that of EC-2. This low potency value was remedied by reconstituting EC-4 with half as much diluent as EC-2. Several laboratories had experienced problems with EC-4 because, with no filler or stabilizer, the vials not only appear empty but it was also quite difficult to dissolve all the endotoxin into the water used for reconstitution. This lot was depleted early in 1981.

Each new lot of endotoxin created problems in that the in-house standards had to be revalidated to reflect the new endotoxin. Therefore, it was decided that in order to avoid the necessity of recalibrating each year or two, a large lot of endotoxin would be prepared that would last 10 to 15 years and could also be offered to the World Health Organization (WHO) as an international standard.

This large lot of endotoxin (Lot EC-5) was prepared from the EC bulk powder. Since Lot EC-2 was probably the best we had prepared, it was decided to prepare Lot EC-5 in the same way. We contracted with a licensed LAL manufacturer that could fill and freeze-dry between 28,000 and 30,000 vials from a single bulk. The protocol used for Lot EC-5 was prepared jointly by the USP and the BOB (now CBER), using lactose and polyethylene glycol as stabilizers.[9]

This lot is stored at 5°C at the USP facility in Rockville, MD and is distributed by the agency as Endotoxin Standard Lot F. We store several hundred vials at the same temperature and distribute it upon request to licensed manufacturers as the U.S. Reference Endotoxin, Lot EC-5. Over the past 6 years approximately 10,000 vials of EC-5 and Lot F (combined) have been distributed. This pattern of use suggests that this endotoxin preparation should last for another 8 to 10 years.

Lot EC-5 is by far the best endotoxin that we have prepared. The only complaints have concerned low potency, and these are rare. If all users would vortex the contents of each vial for 20 min after rehydration with water (as directed), even these few complaints might disappear.

E. FEDERAL REGISTER NOTICES LEADING TO THE LAL GUIDELINE

Since the 1973 **Federal Register Notice**,[12] LAL has proved to be a sensitive indicator of the presence of bacterial endotoxins (pyrogens). Because of this sensitivity, LAL can be of value in preventing the administration of products that may produce fever, shock, and even death in humans or animals if bacterial endotoxins are present.

When the first **Federal Register Notice**[3] was published, experience and available data with LAL were not adequate to support its adoption as the final pyrogen test in place of the rabbit test which had long been recognized as the official test for pyrogen. To make it possible to establish a data base and gain experience with the use of LAL, that notice permitted the introduction of LAL into the workplace without a license from the BOB. Two conditions of this exemption were that its use be limited to the in-process testing of drugs and other products and that the labeling on the LAL container state that the test was not suitable as a replacement for the rabbit pyrogen test.

Since the time of the first notice, production techniques have been greatly improved and standardized so that lysate can be produced consistently to yield an endotoxin sensitivity more than 100 times greater than originally obtained. It is also recognized that the lysate test is faster, more economical, and requires a smaller volume of product than does the rabbit pyrogen test. In addition, the lysate test is less labor intensive than the rabbit test, making it possible to perform many assays in a single day.

In the **Federal Register Notice** of November 4, 1977[13] (42 FR 57749), the FDA defined conditions for the use of LAL as an end product test for endotoxin in human biological products and medical devices. The notice also stated that the application of LAL testing to human drug products would be the subject of a future **Federal Register Notice**.

In March of 1979, the Bureau of Medical Devices, now the FDA Center for Devices and Radiologic Health (CDRH), issued an in-house guideline that recommended procedures for the use of LAL testing as an end product endotoxin test. As a direct result of the CDRH experience in approving petitions for the use of the LAL test in place of the rabbit pyrogen test, several procedures for using the LAL test have evolved and have been adopted for medical devices.

In the **Federal Register Notice** of January 18, 1980[14] (45FR 3668), the FDA announced the availability of a draft guideline that set the formal FDA position on the use of the LAL test as the final pyrogen test for injectable drugs for human and veterinary use, other than licensed biologicals. The agency believed that manufacturers, especially those that have used the LAL test in conjunction with the rabbit pyrogen test, may have had information that could be helpful in preparing the final guideline. After receiving comments and responding to them, a final guideline would be promulgated.

While waiting for public comment, the FDA published additional standards for LAL in the **Federal Register** of May 16, 1980.[15] This was in response to comments to an earlier **Federal Register Notice**, August 11, 1978,[16] that could be helpful in preparing the final guideline.

In response to comments received on the January 18 draft guideline, the FDA made several significant changes (i.e., the deletion of the section, "Absence of Non-endotoxin Pyrogenic Substances", the endotoxin limits change) and many minor editorial changes. The agency also determined that a single document covering all FDA regulated products that may be subject to LAL testing should be made available. Primarily because of the addition of biological products and medical devices to the guidelines, the agency published, in the **Federal Register** of March 29,1983 (45FR 13096), another draft of the guideline available for public comment.[17]

Based on the comments received on the March 29 draft guideline, it was evident that after over a decade of research and **Federal Register** comment, most of the critical issues had been resolved with both the drug industry and USP.[18] Many of the unresolved issues centered around the details of the test applications and the establishment of limits for parenteral products. These issues were incorporated into the final guideline that was published in 1987.[19] Briefly, the significant changes were

1. Requirements to test the sensitivity of a rabbit pyrogen testing colony have been eliminated.
2. Any technique (gel-clot, chromogenic or turbidimetric) can be used in testing a product for endotoxin. However, if a gel-clot lysate is used in a different technique, the results must be interpreted using the criteria for the technique being used.
3. Validation criteria for the chromogenic, endpoint-turbidimetric, and kinetic-turbidimetric LAL technique are included.
4. The CDRH has adopted the USP Endotoxin Reference Standard and revised the limit expressions from ng/ml to EU/ml. This new limit for medical devices is 0.5 EU/ml, except for devices in contact with cerebrospinal fluid (CSF) for which the limit is 0.06 EU/ml. These limits for devices are equivalent to those for drugs for a 70 kg man when consideration is given to the following:
   a. Published FDA studies indicate that less than half added water rinse.

TABLE 1
U.S. Licensed Manufacturers of *Limulus* Amoebocyte Lysate (LAL)

1. Associates of Cape Cod, Inc.
   P. O. Box 224
   Woods Hole, MA 02543

2. Endosafe, Inc.
   1023 Wappoo Road
   Suite 43-B
   Charleston, SC 29407

3. Haemachem, Inc.
   2335 South Handley Blvd.
   St. Louis, MO 63144

4. Neumann Biotechnologies, Inc.
   601 N. Main Street
   Pleasantville, NJ 08332

5. Travenol Laboratories, Inc.
   Hyland Therapeutics Division
   444 West Glenoaks Blvd.
   Glendale, CA 91202

6. Whittaker Bioproducts, Inc.
   Biggs Ford Road
   Walkersville, MD 21793

b. In the worst case situation, all endotoxin present in the confined rinsing of 10 devices could have come from just one device. A wide variation in bioburden is common to some devices.

5. The Center for Drug Evaluation and Research (CDER) and CBER have added a listing of the maximum doses per kilogram per hour and the corresponding endotoxin limits for most of the aqueous injectable drugs and biologics currently on the market. This listing was added to promote uniformity among companies making the same product.

F. LEGAL EFFECT OF THE GUIDELINE

The final guideline was published in December 1987[19] under Section 21 CFR 10.90(b) of the FDA administrative regulations. These regulations provide for use of guidelines that are acceptable to the FDA for subject matter within its statutory authority. The guidelines are *not* legal requirements. However, a person who follows the guidelines may be assured that the procedures will be acceptable to the FDA. The LAL guidelines were developed to inform manufacturers of human drugs, biologics, animal drugs, and medical devices of procedures that the FDA considers necessary to validate the use of LAL as an end product endotoxin test.

G. LAL LICENSED MANUFACTURERS

The six licensed LAL manufacturers are listed in Table 1. They are all licensed for the gel test. Whittaker Bioproducts, Inc. is also licensed for the chromogenic test, and Associates of Cape Cod has been approved for a turbidimetric test. Lysates of both Marine Biologicals and Associates of Cape Cod have been approved for testing CSF for endotoxin. The entire production capacity of Travenol Laboratories is used for testing within the corporation.

II. METHODS APPROVED BY FDA FOR THE LAL TEST

A. GEL END POINT METHOD

For many years the gel test was the only method for the LAL test. The details have been described many times.[20-24] Essentially, the test consists of adding 0.1 ml of lysate to 0.1 ml of product to be tested in a 10 × 75 mm pyrogen-free test tube, placing it in a 37°C water bath for 1 h and then carefully inverting the tube 180°. If the solution has formed a gel and remains in the bottom of the tube, the test is positive. Conversely, if the solution remains fluid, the test is negative. If the mixture is turbid or contains granules and does not remain in the bottom of the tube when inverted, the test is still negative. There are no plus/minus readings. The lysate sensitivity is determined by testing an unknown lysate with the U.S. Standard Endotoxin EC-5 in comparison to the U.S. Reference Lysate as described earlier (Section I.C).

B. CHROMOGENIC SUBSTRATE METHOD

There is one manufacturer (Whittaker Bioproducts, Inc.) licensed for producing the chromogenic method of performing the LAL test. This method utilizes a modified LAL and a synthetic color-producing substrate to detect endotoxin. The chromogenic LAL test is a quantitative test for Gram-negative bacterial endotoxin.[25-27] An unknown sample is mixed with the lysate supplied in the test kit and incubated at 37°C for 10 min. A substrate solution is then mixed with the lysate sample and incubated at 37°C for an additional 3 min. While still in the water bath, the reaction is stopped with 25% acetic acid. Once the reaction is stopped, the final reading can be done in a spectrophotometer at any time. If endotoxin is present in the sample, a yellow color will develop. The absorbance of the sample is determined spectrophotometrically at 405 nm. Since the absorbance is in direct proportion to the amount of endotoxin present, the concentration of endotoxin can be calculated from a standard curve. Such a procedure is amenable to automation and is useful when making numerous determinations.

C. TURBIDIMETRIC METHOD

There is one manufacturer (Associates of Cape Cod, Inc.) licensed for the turbidimetric method for performing the LAL test. This method, like the chromogenic, requires a spectrophotometer to measure the change in optical density.[28] Using the turbidimetric method, one measures turbidity due to the reactions between the LAL and endotoxin.[29,30] A positive test is determined by the formation of visible turbidity which is measured (by visible light) as an increase in absorbance. The increase in absorbance is directly related to endotoxin concentration. The endotoxin concentration can be determined since the data can be plotted in a linear form and compared to a standard curve.

III. AUTOMATED SYSTEMS APPROVED BY FDA

A. THE LAL-5000 AND THE KINETIC TURBIDIMETRIC ASSAY

The kinetics of the LAL reaction were first described by Levin and Bang.[31] Basically, the formation of a gel-clot is dependent on time, temperature, endotoxin concentration, and the concentration of proclotting enzyme (LAL) and is preceded by the development of turbidity. Kinetics may be obtained by measuring the turbidity as a function of time. In the turbidimetric method, temperature and LAL concentration are fixed, and a standard curve is constructed using known concentrations of endotoxin as well as the time taken for each to reach a given level of turbidity.

The LAL-5000 is an incubating optical reader with an associated computer. It has been tested and approved for determining endotoxin by the kinetic turbidimetric method in samples with lots of Associates of Cape Cod lysate labeled "GT". The "GT" label indicates that the same lysate can be used for the gel and kinetic turbidimetric tests. The LAL-5000 has a detection range of 0.001 to 100 EU/ml.

The kinetic turbidimetric assay (using visible light) is based upon the end point turbidimetric method. The assay depends upon the principle that the rate of turbidity development in an LAL/sample mixture increases with endotoxin concentration. In the case of the LAL-5000, the time taken to reach a particular optical density threshold is determined. This time is called *onset time*. The onset time decreases with higher endotoxin concentrations in the samples. The log of the endotoxin concentration is inversely proportional to the log of the onset time.

The advantages of the kinetic turbidimetric assay are convenience, simplicity (given computerized equipment), and both quantitative and qualitative information about the LAL/endotoxin reaction in the sample.

B. THE TOXINOMETER ET-201

This is a precise, efficient, automated method for measuring turbidity change in the gel-clot endotoxin reaction.[32,33] This instrument monitors the ratio R (t) of the sequential to the initial transmittance of up to 64 samples simultaneously and independently at 12 s increments at a wavelength of 660 nm and quantifies endotoxin by determining the gelation time of the sample defined as the time required to obtain a 5% decrease of R (t).

The well-correlated standard curve was obtained for endotoxin concentrations from 0.0005 to 500 EU/ml. The detection of endotoxin obtained by this method was in good agreement with that obtained by the conventional gel-clot method. In fact, the titration read by the ET-201 after about 30 min of incubation is the same titration read by the gel test in 60 min. The toxinometer ET-201 method is very useful for the quantitative determination of endotoxin in routine clinical and pharmaceutical samples. The advantages of this method are

1. The turbidimetric and gel-clot are read in the same tube.
2. All gel-clot lysates can be used.
3. Enhancement or inhibition can be measured by the alteration in the gelation time.
4. Many determinations can be made in a short period of time.

IV. LABELING THE LAL FOR TESTING BODY FLUIDS FOR ENDOTOXIN

If a licensed LAL manufacturer requests approval to label the product for detecting endotoxin in a body fluid, a product license amendment must be submitted. This amendment must be accompanied by data supporting the amendment request and a copy of the proposed labeling. The amendment is then reviewed by a committee which either accepts or rejects the proposal. The manufacturer is then notified of the decision. Over the past 8 years, we have approved two such amendments, one for detecting endotoxin in spinal fluid[34-37] and the other for detecting gonorrhea in males.[38,39]

The clinical utility of the LAL test depends upon specific clinical situations. In diagnosing Gram-negative septicemia or Gram-negative pyogenic arthritis, the utility of the LAL test is still unreliable. A sample of this type (blood or joint fluid) has to be altered in some way so that the inhibitors are neutralized, allowing the endotoxin to be detected by the LAL test. The treatment with chloroform and boiling have removed some of the inhibitors but studies have shown that even with this treatment, the results of testing such samples are not reliable enough to obtain FDA approval for making a diagnosis with the LAL test.

The LAL test is clearly useful in the clinical management of meningitis, because there are very few inhibitors in CSF and the test can rapidly detect endotoxin with a high degree of sensitivity and greater than 98% specificity. The FDA has approved the LAL test for diagnosing meningitis.

Since *Neisseria* are Gram-negative bacteria and produce an abundance of endotoxin it is not surprising that the LAL test is extremely useful and specific in diagnosing gonorrhea in males. The same is not the case in females, since the mixed bacterial flora causes the LAL test to be much less specific. The FDA has approved the LAL test as a rapid presumptive diagnosis of gonorrhea in males, but the lysate used for approval is no longer available. More data are needed before a female diagnosis can be approved.

The pharmaceutical community is still hesitant about relying definitively on the LAL test for the detection of endotoxin. It is usually used in conjunction with the conventional rabbit pyrogen test so that data can be accumulated to assess the advantages and limitations of this relatively new, rapid, sensitive, and specific test for bacterial endotoxin.

V. TESTING U.S. STANDARD ENDOTOXIN (LOT EC-5) IN RABBITS AND HUMANS

While the LAL test was used to set allowable endotoxin limits for biologicals, drugs, and devices, the major frame of reference was rabbit data with Lot EC-5.[10] Results indicated that greater than 50% of the rabbits receiving 10 EU/kg will show a temperature rise of 0.5°C or higher while 65% of the rabbits receiving 20 EU/kg will have at least a 0.5°C rise. However, there are several conflicting reports in the literature as to how rabbits and humans differ in their sensitivity to endotoxin.

On one hand, Dare and Mogey[40] found that the rabbit is 0.33 to 7.0 times as sensitive to endotoxin as humans, depending on experimental conditions. On the other hand, Wolff[41] states that, for the same febrile response in rabbits and humans, the rabbit requires 10 times more endotoxin than man on a ug/kg basis. Several years earlier, Greisman and Hornick[42] found that the threshold pyrogenic dose of endotoxin on the basis of weight was the same for both species.

Since samples of this material (EC-5) are available to pharmaceutical manufacturers and selected biomedical investigators as an aid to the standardization of bioassays and research with control standard endotoxin, it is important to define the properties of this endotoxin in humans.

Elin et al.,[43] and Csako et al.,[44] tested the stock EC endotoxin (0.035, 0.1, 0.35, 0.5, 1, 2, 3, and 4 ng/kg) which was administered intravenously to groups of four normal volunteers. The temperature was recorded every 30 min for 9 h. The results indicated that the minimal pyrogenic dose was probably somewhere between 0.1 and 0.5 ng/kg. This study did not satisfy our need to define the U.S. Standard Endotoxin in man because it utilized a stock EC and not EC-5 and it utilized weight (ng) as an endotoxin value and not the endotoxin units that define the standard.

Since we release products for human use based on EU, it was imperative that we test endotoxin units in man. Therefore, in April 1988, we contracted with the Clinical Research Center of the Tulane Medical School to determine the minimal pyrogenic dose of EC-5 endotoxin in man.

The Clinical Research Center study involved 60 healthy male subjects between the ages of 18 and 45 years. Each subject had a normal, stable body temperature and passed a physical examination. There were groups of 12 volunteers per dose level plus 12 for placebo. The dose levels for this double blind, parallel, randomized placebo control study were 2, 4, 8, and 16 EU/kg along with the placebo of pyrogen-free water that was used to dilute the freeze-dried endotoxin.

The contents of four vials of EC-5 were rehydrated with pyrogen-free water to complete this clinical study. Following rehydration, the contents of each vial were tested for potency using the LAL test. The results indicated that each vial contained 10,000 EU as labeled. This study has been evaluated and the threshold pyrogenic dose was between 4 and 8 EU/kg with the peak temperature rise occurring between hours 4 and 5 following the intravenous injection of the EC-5 endotoxin. This study should be published in detail in 1990.

VI. SUMMARY

The LAL test is not only a rapid, highly specific, reproducible test but it is the most sensitive test available to detect Gram-negative bacterial endotoxins (pyrogens). It is a relatively simple test to perform and extremely cost-effective compared to the official rabbit pyrogen test. The LAL test has been used most successfully in the pharmaceutical industry where rapid and economical testing of medical devices, biologicals, and parenteral drugs for endotoxin contamination is critical. In fact, it is used quite extensively for in-house monitoring of each step in the preparation of numerous parenteral fluids. Now that there are clinical data available indicating the dose of EC-5 necessary to cause a minimal pyrogenic response, more manufacturers may feel comfortable releasing final container material using the LAL test.

There are now six licensed LAL manufacturers, each with a capacity of producing lysate with a sensitivity as low as 0.03 EU/ml, which is as much sensitivity as needed in today's scientific community. The lysate produced for automated systems (chromogenic and turbidimetric) may be more sensitive, but the main advantage of automation is rapid test results with more samples being processed in a shorter period of time.

REFERENCES

1. **Hochstein, H. D., Seligmann, E. B., Marquina, R. E., and Rivera, E.,** Limulus amebocyte lysate testing of Normal Serum Albuman (Human) released in the United States since 1975, *Dev. Biol. Stand.*, 44, 35, 1979.
2. **Hochstein, H. D.,** The LAL test versus the rabbit pyrogen test for endotoxin detection, *Pharm. Tech.*, 36, 1981.
3. Federal Register Notice, Biological Products, Status of Biological Substances Used for Detecting Bacterial Endotoxins (38 FR 1404), January 12, 1973.
4. **Rastogi, S. C., Seligmann, E. B., Hochstein, H. D., Dawson, J. H., Farag, L. G., and Marquina, R. E.,** Statistical procedures for evaluating the sensitivity of Limulus amebocyte lysate by using a reference lysate, *Appl. Environ. Microbiol.*, 38, 911, 1979.
5. Code of Federal Regulations, Additional Standards for Limulus Amebocyte Lysate — Final Rule, Title 21, parts 610 and 660.100, 1980.
6. **Rastogi, S. C., Hochstein, H. D., and Seligmann, E. B., II,** Quality control procedure for testing Limulus amebocyte lysate sensitivity using reference endotoxin 1b, *Bull. Parent. Drug Assoc.*, 31, 120, 1977.
7. **Castor, G. B., Kantor, N., Blakely, J., Nielsen, J. K., Randolph, J. O., and Kirshbaum, A.,** Characteristics of a highly purified pyrogenic lipopolysaccharide, *J. Pharm. Sci.*, 60, 1578, 1971.
8. **Rudbach, J. A., Akiya, E. I., Elin, R. J., Hochstein, H. D., Luoma, E. C. B., Milner, C. B., Milner, K. C., and Thomas, K. R.,** Preparation and properties of a national reference endotoxin, *J. Clin. Microbiol.*, 3, 21, 1976.
9. **Hochstein, H. D.,** U.S. government quality control program for Limulus amebocyte lysate and endotoxin, in *Bacterial Endotoxin: Structure, Biomedical Significance, and Detection with Limulus Amebocyte Lysate Test*, Alan R. Liss, New York, 1985, 221.
10. **Hochstein H. D., Mills, D. F., Outschoom, A. S., and Rastogi, S. C.,** The processing and collaborative assay of a reference endotoxin, *J. Biol. Stand.*, 11, 251, 1983.
11. Expert Committee on Biological Standardization, WHO Tech Rep. Ser., 147:11, 1958; WHO unpublished working document no. WHO/BS/400.
12. Federal Register Notice, Limulus Amebocyte Lysate (Additional Standards) (38 FR 180), September 18, 1973.
13. Federal Register Notice, Licensing of Limulus Amebocyte Lysate. Use as an Alternative for Rabbit Pyrogen Test, (42 FR 213), November 4, 1977.
14. Federal Register Notice, Human and Veterinary Drugs; Availability of Draft Guideline for Use of Limulus Amebocyte Lysate (45 FR 13), January 18, 1980.
15. Federal Register Notice, Human, Biological, and Animal Drugs and Medical Devices; Availability of Draft Guideline for Use of the Limulus Amebocyte Lysate (LAL) Test (148 FR 61), March 29, 1983.
16. Federal Register Notice, Additional Standards for Diagnostic Substances for Laboratory Tests, (45 FR 97), May 16, 1980.
17. Federal Register Notice, Limulus Amebocyte Lysate, Proposed Additional Standards (43 FR 156), August 11, 1978.
18. *The United States Pharmacopeia*, Vol. 20, Mack Publishing, Easton, PA, 1980, 888.

19. Guideline on Validation of the Limulus Amebocyte Lysate Test as an End-Product Endotoxin Test for Human and Animal Parenteral Drugs, Biological Products, and Medical Devices, U.S. Department of Health and Human Services, December 1987.
20. **Mascoli, C. C. and Weary, M. E.,** Limulus amebocyte lysate (LAL) test for detecting pyrogens in parenteral injectable products and medical devices; advantages to manufacturers and regulatory officials, *J. Parenteral Drug Assoc.*, 33, 81, 1979.
21. **Elin, R. J.,** Clinical utility of the Limulus test with blood, CSF and synovial fluid, in *Biomedical Applications of the Horseshoe Crab (Limulidae)*, Cohen, E., Ed., Alan R. Liss, New York, 1979, 279.
22. **Cooper, J. F., Hochstein, H. D., and Seligmann, E. B.,** The limulus test for endotoxin (pyrogen) in radiopharmaceuticals and biologics, *Bull. Parent. Drug Assoc.*, 26, 153, 1972.
23. **Prior, R. B.,** Bacterial endotoxin detection by the *Limulus* amebocyte lysate test, in *Antigen Detection to Diagnose Bacterial Infections*, Vol. 2, Kohler, R. B., Ed., CRC Press, Boca Raton, FL, 1986, 242.
24. **Jorgensen, J. H. and Smith, R. F.,** Preparation, sensitivity, and specificity of Limulus lysate for endotoxin assay, *Appl. Microbiol.*, 26, 43, 1973.
25. **Scully, M. F., Newman, Y. M., Clark, S. E., and Kakkar, V. V.,** Evaluation of a chromogenic method for endotoxin measurement, *Thromb. Res.*, 20, 263, 1980.
26. **Nachum, R. and Berzofsky, R. N.,** Chromogenic *Limulus* lysate assay for rapid detection of gram negative bacteriuria, *J. Clin. Microbiol.*, 21, 759, 1985.
27. **Prior, R. B. and Spagna, V. A.,** Rapid evaluation of gonococcal and nongonoccal urethritis in men with *Limulus* amebocyte lysate and a chromogenic substrate, *J. Clin. Microbiol.*, 17, 485, 1983.
28. **Hollander, V. P. and Harding, W. C.,** A sensitive spectrophotometric method for measurement of plasma endotoxin, *Biochem. Med.*, 15, 28, 1976.
29. **Jorgensen, J. H. and Alexander, G. A.,** Automation of the *Limulus* amebocyte lysate test by using the Abbott MS-2 microbiology system, *Appl. Environ. Microbiol.*, 41, 1316, 1981.
30. **Novitsky, R. J., Roslansky, P. F., Siber, G. R., and Warren, H. S.,** Turbidimetric method for quantifying serum inhibition of *Limulus* amebocyte lysate, *J. Clin. Microbiol.*, 20, 211, 1985.
31. **Levin, J. and Bang, F. B.,** Clottable protein in Limulus: its localization and kinetics of its coagulation by endotoxin, *Thromb. Diath. Haemorrh.*, 19, 186, 1968.
32. **Oishi, H., Takaoka, A., Hatayama, Y., Matsuo, T., and Sakata, Y.,** Automated Limulus Amebocyte Lysate (LAL) Test for Endotoxin Analysis using a New Toxinometer ET-201, *J. Parent. Sci. Tech.*, 39, 194, 1985.
33. **Oishi, H., Halayama, Y., Siraishi, H., Yanagisawa, K., and Sakata, Y.,** The automated measurement of endotoxin by parallel turbidimetric time assay, *Yakugaku Zasshi*, 105, 300, 1985 (in Japanese).
34. **Berman, N. S., Siegel, S. E., Nachum, R., Lipsey, A., and Leedom, J.,** Cerebrospinal fluid endotoxin concentrations in gram-negative bacterial meningitis, *J. Pediatr.*, 88, 553, 1976.
35. **Jorgensen, J. H. and Lee, J. C.,** Rapid diagnosis of gram-negative bacterial meningitis by the *Limulus* endotoxin assay, *J. Clin. Microbiol.*, 7, 12, 1978.
36. **Dyson, D. and Cassady, G.,** Use of Limulus lysate for detecting gram-negative neonatal meningitis, *Pediatrics*, 58, 105, 1976.
37. **McCraken, G. H. and Sarff, L. D.,** Endotoxin in cerebrospinal fluid: detection on neonates with bacterial meningitis, *JAMA*, 235, 617, 1976.
38. **Spagna, V. A., Prior, R. B., and Perkins, R. L.,** Rapid presumptive diagnosis of gonococcal urethritis in men by the limulus lysate test, *Br. J. Vener. Dis.*, 55, 179, 1979.
39. **Chapel, T. A., Adcock, M., Smith, B., Barnes, T., and Goodman, M.,** Evaluation of the *Limulus* amebocyte lysate assay for presumptive diagnosis of gonorrhea in men at a clinic for sexually transmitted diseases, *Sex. Transm. Dis.*, 8, 175, 1981.
40. **Dare, J. G. and Mogey, G. A.,** Rabbit responses to human threshold doses of bacterial pyrogen, *J. Pharm. Pharmacol.*, 6(5), 325, 1954.
41. **Wolff, S. M.,** Biological effects of bacterial endotoxins in man, in *Bacterial Lipopolysaccharides: The Chemistry, Biology, and Clinical Significance of Endotoxins*, Kass, E. H. and Wolff, S. M., Eds., University of Chicago Press, Chicago, 1973, 251.
42. **Greisman, S. E. and Hornick, R. B.,** Comparative pyrogenic reactivity of rabbit and man to bacterial endotoxin, *Proc. Soc. Exp. Biol. Med.*, 131, 1154, 1969.
43. **Elin, R. J., Wolff, S. M., McAdam, K. P. W. J., Cedid, L., Audibert, F., Bernard, C., and Oberling, F.,** Properties of Reference Escherichia coli endotoxin and its phthalylated derivative in Humans, *J. Infect. Dis.*, 144, 329, 1981.
44. **Csako, G., Elin, R. J., Hochstein, H. D., and Chao-Ming Tsai,** Physical and biological properties of U.S. Standard Endotoxin EC after exposure to ionizing radiation, *Infect. Immun.*, 41, 190, 1983.

Chapter 5

DETECTION OF ENDOTOXEMIA

Frederick C. Pearson

TABLE OF CONTENTS

I. Overview 52

II. Clinical Syndrome 53

III. Clinical Diagnosis 55

IV. Laboratory Diagnosis 55

V. Sample Preparation 56
- A. False Negatives 56
- B. False Positives 57
- C. Heat and Dilution 57
- D. Platelet Extraction 58
- E. Ammonium Sulfate Precipitation 58
- F. Ether Method 58
- G. Perchloric Acid 59
- H. Resin Extraction 59
- I. Chloroform Extraction 59
- J. pH Shift Method 59

VI. Current Status of the LAL Test 60

VII. Future Clinical Trend 61

References 62

I. OVERVIEW

Endotoxemia as a clinical entity in the human is best described as the collective physiologic responses of various organ systems to bacterial toxic substances known as endotoxins. Organ system dysfunction occurs even in the absence of Gram-negative bacterial invasion of a specific organ. These endotoxins are derived from the constituents found within the outer membrane of Gram-negative bacteria. Disruption of this membrane either releases or exposes active sites on endotoxin. In its purified state, endotoxin is known as lipopolysaccharide (LPS). A portion known as lipid A is believed to be the active moiety.[1,2]

Several clinical situations present frequently as a result of endotoxin elaboration within the circulatory system. Any portion of the body exposed to Gram-negative bacteria is capable of barrier dysfunction resulting either in the release of endotoxin into the systemic circulation or the colonization of these bacteria within a localized abscess or in widespread distribution. Examples of barrier dysfunction include Gram-negative septic shock as a complication of burns of the skin, neonatal sepsis, and enteritis.

Endotoxin confined within the outer membrane of Gram-negative bacteria confers little toxicity per se compared to when release is affected by cell disruption. In the natural state, whereby specific components of the immune system phagocytize Gram-negative bacteria while others neutralize free endotoxin, this process occurs at a rate-limited progression. When medical intervention with intravenous (IV) antibiotics hastens the disruption of these bacteria, a rapid rise in the shock state becomes evident from the liberation of endotoxin.

Shenep et al.[3] evaluated this relationship by measuring endotoxin levels in plasma both pre- and postantibiotic therapy. Markedly increased endotoxin levels became evident as bacteremia subsided. They also theorized from their studies that Gram-positive infection could result in endotoxemia but by overwhelming the reticuloendothelial system to allow endotoxin from intestinal flora to escape into the systemic circulation by way of the hepatoportal circulation.

Gram-negative bacteremia is a common infection in burn patients because of the breakdown in their protective skin barrier. Colonization of exposed subcutaneous and necrotic dead tissues embedded in a serosanguinous solution is difficult to prevent and has led to early debridement of burn wounds in recent years. The greater the percent of area burned, the greater the mortality from the effects of endotoxin on the body organ systems.

In some instances, compromise of the liver macrophage system (Kupffer cells) from hepatic insufficiency — regardless of the cause — will lead to improper or inadequate detoxification of endotoxin from portal system drainage and result in endotoxemia. This state is noted with some degree of frequency within the newborn, whose liver and reticuloendothelial system function are still immature, and in those infants with necrotizing enterocolitis,[4] wherein gastrointestinal flora may leak into the systemic circulation. Documented evidence of "intestinal endotoxin", whereby the endotoxin produced by endogenous Gram-negative bacteria reaches the systemic circulation has been reported.[5] The Gram-negative pathogens most frequently isolated and shown responsible for septic shock are *Escherichia coli*, *Klebsiella pneumoniae*, and *Proteus* sp.[6] Endotoxic shock occurs in about 25% of patients infected by Gram-negative bacteria. The condition has a reported mortality rate of 60 to 70%.

Endotoxin itself has a relatively simple molecular structure that exists most commonly in a bilayer form and has been shown to consist of lipid A, a core polysaccharide, and a serospecific polysaccharide side chain component. Lipid A is covalently bound to the polysaccharide core and is thought to mediate the major biological effects of endotoxins.[7] Recently, the ability of endotoxin to induce the production of interleukin-1 (IL-1) and tumor necrosis factor has been noted.[8,9] Other physiologically active factors found following endotoxin stimulation include complement[10] and both bradykinin and vasoactive kinin.[11] Endotoxin also frequently stimulates the intrinsic coagulation mechanisms, resulting in disseminated intravascular coagulation (DIC) syndrome.[12,13]

Common portals of Gram-negative bacterial entry include the umbilical cord in the newborn, burns of the skin, the urinary tract system, and — the greatest reservoir — the gastrointestinal tract. Contamination from the gastrointestinal tract often results from trauma, surgical manipulation, and degenerative conditions due to aging and long- and short-term effects of medication and drugs.

Bacterial endotoxins do not exhibit a direct effect on organs and tissues of the body, but act through the disruption of normal organ system function. Although an initial rise in blood pressure may be observed, a secondary hypotensive reaction ensues and leads to acidosis and hypoxia.

Septic shock is a medical emergency that requires immediate attention with antibiotic therapy, reversal of hypotension, and management of widespread organ dysfunction. Recent evidence tends to indicate an interrelationship between the immune system and endotoxemia.[14]

Cells of the monocyte/tissue macrophage lineage secrete a variety of soluble immune system mediators collectively known as cytokines.[15] One cytokine, tumor necrosis factor, has been considered to be a prime mediator of the response of the body to Gram-negative bacterial sepsis.[16] Tumor necrosis factor was detected, after lethal endotoxin infusion, in the sera of experimental animals.[17] In healthy human male volunteers, a pulse of tumor necrosis factor was noted following endotoxin administration.[18] The dose utilized, however, was insufficient to produce hypotension.

Tumor necrosis factor induces the release of other products, such as IL-1, from monocytes and endothelial cells[19] and promotes blood coagulation by increasing procoagulant activity. Endothelial cell growth and adhesion to inflammatory cells are also enhanced.[20] On the basis of Ibuprofen® pretreatment and the observed secondary amelioration of symptoms, the cyclooxygenase system was felt to be the effector pathway.[18] Ibuprofen®, however, does not prevent the production of tumor necrosis factor.

Other contributory evidence has been developed. Passive immunization by means of antisera has been noted to be protective against endotoxemia,[21] as has monoclonal antibody treatment.[22,23] Human monoclonal IgM antibody to lipid A (core LPS) administered 24 h after endotoxin challenge also offered protection from the effects of *Pseudomonas* endotoxin.[24] In another study, antibody to tumor necrosis factor protected against hypotension, fibrin deposition, and death in a rabbit model of endotoxin shock.[25] Tumor necrosis factor level peaks in the systemic circulation at 90 min after endotoxin release and lasts for 3 to 4 h.[18] Other cytokines measured — IL-1 and β- and γ-interferon — were unchanged in this study although previous reports by Sieff et al.[26] indicated IL-1 production. They further noted that detectable tumor necrosis factor occurred only at the beginning of clinical responses.

II. CLINICAL SYNDROME

Septic shock is a clinical syndrome of multiple organ system failure and secondary hypotension that results from the presence of either Gram-negative bacteria or endotoxin in the systemic circulation. The source of the bacteria has been observed to be the single most important factor with regard to mortality.[6] Lethal infection was associated most often with bacterial colonization of the respiratory or gastrointestinal tract. Any compromising factors such as age, coincident organ-system disease, recent surgery, and neoplastic disease alter recovery capability and increase the risk of death.

The most commonly observed symptoms in Gram-negative sepsis are fever, chills, hyperventilation, altered mentation, and variation in hemodynamic status most often resulting in hypotension and cardiovascular collapse.[6] Two main phases occur. First, profound vasodilation becomes evident with increased cardiac output and lowered peripheral resistance. Second, a marked and extensive capillary-leak syndrome results in hypovolemia.[27] Often irreversible shock ensues. The capillary-leak syndrome has been considered to be a result of widespread complement activation.[28]

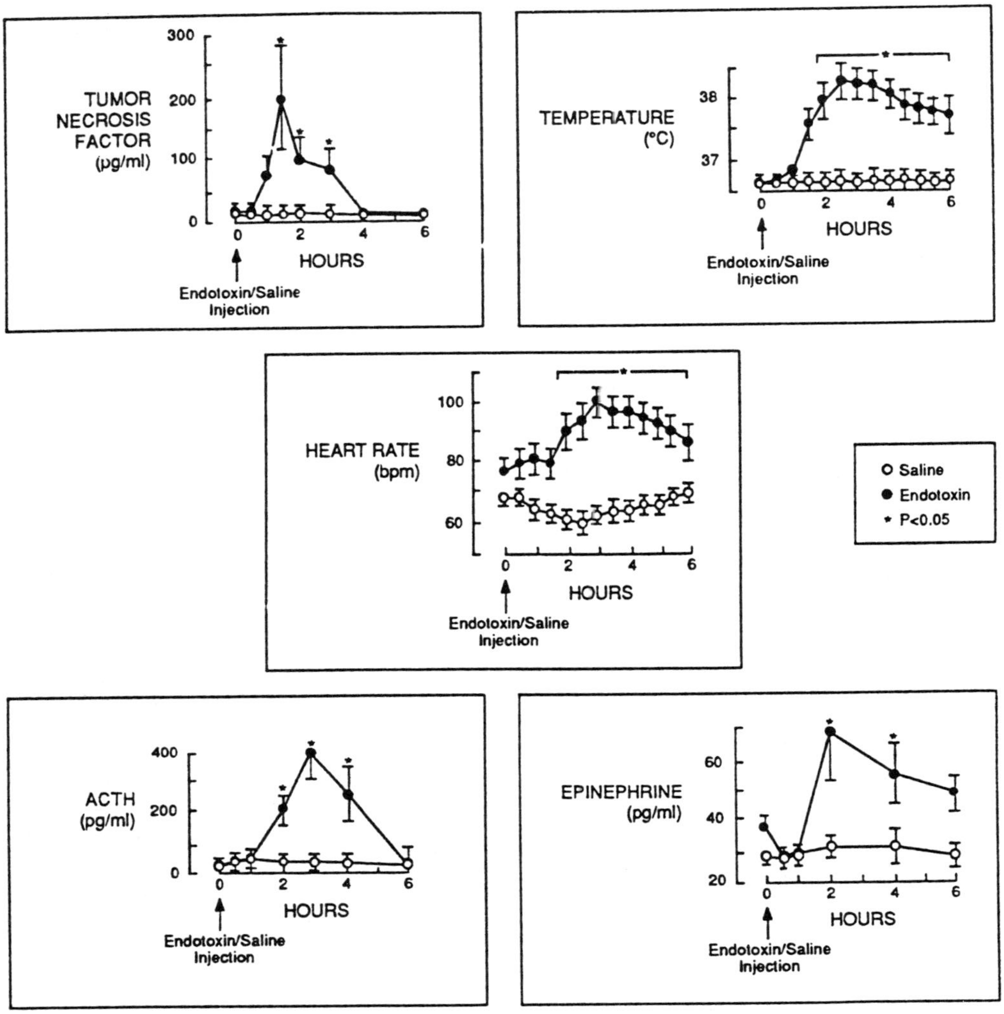

FIGURE 1. Alterations in concentrations of circulating tumor necrosis factor in temporal relation to certain physiologic and neurohumoral changes in six subjects from whom samples were obtained at frequent intervals after endotoxin and saline injection. (From Michie, H. R., Manogue, K. R., Spriggs, D. R., Revhaug, A., O'Dwyer, S., Dinarello, C. A., Cerami, A., Wolff, S. M., and Wilmore, D. W., *N. Engl. J. Med.*, 318, 1481, 1988. Reprinted with permission from *The New England Journal of Medicine*.)

DIC is often observed and has been long considered to be related to LPS-platelet[29] or Hageman factor[30,31] interaction of the intrinsic clotting mechanism.

The recent study by Michie et al.[18] of human volunteers administered purified endotoxin infusions, was a significant addition to the understanding of the effects of low-dose endotoxin (Figure 1). At this dose (4 ng/kg, *E. coli* endotoxin) no significant change in either IL-1 or γ-interferon could be detected at any time after endotoxin or saline control infusions. They could not, however, utilize endotoxin dose ranges that would precipitate the full spectrum of Gram-negative sepsis, including hypotensive reactions. Aulitzky et al.,[32] in an investigation of the effects of tumor necrosis factor infusion in 12 patients with advanced malignant disease, found equivocal responses with regard to γ-interferon production, but felt overall that some production has been induced.

III. CLINICAL DIAGNOSIS

The initial diagnosis of sepsis is one of clinical suspicion by the attending physician. Usually the patient exhibits fever, chills, and hypotension and is clearly ill. This presentation is altered in the very young and the very old, and many times an accurate diagnosis is delayed by the paucity of symptoms. In the case of the former, the enzyme systems of the body are immature, while in the latter, they have grown to be sluggishly responsive to endogenous stimuli.

Fever is a standard response to substances released by the activation of monocytes and tissue macrophages. Three cytokines have been observed to act as pyrogens:tumor necrosis factor, IL-1, and α-interferon.[33] All induce the production of prostaglandin E2, which then acts at the hypothalamic level and affects thermogenesis through the cyclooxygenase system mentioned previously.

The hemodynamic alterations are a result of the interplay between compensatory mechanisms of the central nervous system and the cardiovascular system with bradykinin-induced peripheral vasodilatation.[6] Ventilation-perfusion abnormalities of the lungs can lead to adult respiratory distress syndrome (ARDS) and right to left shunting resistant to oxygen therapy.[33] Acidosis and hypoxia result and are often refractory to treatment.

IV. LABORATORY DIAGNOSIS

Gram-negative sepsis typically shifts the differential white blood cell (WBC) count toward the production of polymorphonuclear leukocytes. Continued infection and release of endotoxin stimulates tumor necrosis factor and subsequently granulocyte-monocyte-colony stimulating factor (GM-CSF) and granulocyte-colony stimulating factor (G-CSF). These stem cell inducers maintain the inflammatory response.[34]

Often the laboratory report has notations of neutrophilic changes such as toxic granulations, Dohle bodies (cytoplasmic remnants of an earlier developmental stage), and vacuolization. Right to left shunts in the lung result in characteristic reductions in $p0_2$ in arterial blood as measured by arterial blood gas studies. Fibrin degradation products, prolonged prothrombin times, and reduced fibrinogen levels are also seen in the DIC syndrome. Both hyper- and hypoglycemia have been noted, making this determination less diagnostic for sepsis.

Blood cultures are drawn at the first suspicion of sepsis, but reports often require 3 d to complete and in many cases are negative for growth. A decision regarding antibiotic administration is often made on an empiric basis from clinical parameters.

Other laboratory attempts have been made to make a determination of sepsis on the basis of a relatively shorter time interval. One such study indicated that a determination of C-reactive protein (CRP) in cerebrospinal fluid (CSF) is an accurate indicator capable of differentiation between viral and bacterial etiologies.[35] (Serum CRP is an acute-phase protein that rises rapidly in the presence of bacteremia.) There were 68 consecutive cases of meningitis (39 bacterial and 29 viral) evaluated over a 3-year period. Serum CRP was found to have a sensitivity of 100% and a specificity of 96%. The presence of CRP in CSF had a sensitivity of 94% and a specificity of 93%.

Since the demonstration of a practical assay for endotoxin by means of the *Limulus* amoebocyte lysate (LAL) test by Levin and Bang,[36] more recent attempts to utilize this test system with modifications have been reported (see current technology review in this chapter). The LAL test has achieved some success in testing for the presence of endotoxin in urine, synovial fluid, CSF, and urethral and cervical exudates,[37,38] while detection of endotoxin in blood has led to many alterations in the LAL test.[39-41] No other endotoxin evaluation system offers the medical profession a more rapid, accurate, and reproducible means for early evaluation of Gram-negative bacterial endotoxin generated from infections.

Although the LAL test is considered specific for the presence and detection of endotoxin in many test situations, such as the final pyrogen release of pharmaceuticals and medical devices, other substances, particularly high molecular weight carbohydrates including synthetic dextrans, carrageenan, (1–3)-β:D-glucan, mannans, and dextrans[42] have been observed to react in the test. These false-positive materials do not induce the picture typically seen in septic shock, however, and do not induce the production of IL-1 in *in vitro* systems.[43]

V. SAMPLE PREPARATION

Due to LAL test inhibition and enhancement by human plasma, preparation of the sample is critical to both the sensitivity and the specificity of the test result. Special attention has been paid to the development of methodologies to offset this interaction that often leads to either false positives and false negatives. A review of the methods that have been developed was covered in an earlier work and is summarized in this section.[44]

A. FALSE NEGATIVES

Early attempts to recover known amounts of endotoxin that had been added to human serum resulted in low recovery.[45] Other studies suggested the presence of an inhibitor,[46] which could be negated by either dilution of at least 1:10 or by prior extraction with chloroform over a 1-h time period.

Complement, particularly the C5 and C6 components, has been shown to detoxify endotoxin,[47] but this interaction may be readily eliminated by the heat and dilution method. Complement inactivation occurs at half the time and two thirds the temperature used in a standard heat and dilution procedure discussed here.

Johnson et al.,[48] isolated from human serum a protein with the capacity to irreversibly inactivate lipopolysaccharide. The exact nature of this inhibitor was not determined. At least two α-globulins from human serum have been observed to detoxify lipopolysaccharide.[49-51]

Rosen[52] noted detoxification of endotoxin by a fraction obtained after the use of calcium binding anticoagulants, thereby suggesting that heparin should be utilized in the presence of an adequate amount of divalent cations in order to eliminate heparin inhibitory activity. In additional test systems, any reduction of calcium concentration resulted in lower levels of detection of endotoxin.[48,53,54]

The ancient clotting system of the horseshoe crab can be surmised to be a simple system homologous to the clotting system in human plasma. The *Limulus* clotting enzyme reacts to substrates as does activated factor X,[55] and is neutralized by soybean trypsin inhibitor.[56] The *Limulus* proclotting enzyme has been shown to be a serine protease similar to trypsin.[56] α-1-Antitrypsin and antithrombin III have both been noted to be potent inhibitors of the LAL test[57] and are both eliminated by organic solvents.[58,59]

Circulating immune complexes also have been noted to mask the detection of endotoxin by the LAL test, and may provide the etiology of the different responses noted in humans who have been variably exposed in the past to endotoxin challenge.[60] Chloroform extraction was observed to be poorly effective after immune complexes had formed. Ziegler[61] reported higher antibody titers in endotoxin immunized men in whom preimmune titers were low.

High density lipoproteins (HDLs) have been observed to detoxify endotoxin by means of the formation of HDL-LPS complexes.[51,63,64] Delipidation of serum results in reduced lipopolysaccharide pyrogenicity that returns after HDL reconstitution. HDL may act to reduce LPS activity by a similar clearing action that reduces cholesterol esters.

Platelets and Factor XII are involved in the intrinsic coagulation system. Platelets have been shown to be associated with LPS following combination with plasma.[64,65] Factor XII has been shown to be activated by the infusion of endotoxin.[66] A component of the immune system appears to be involved in the human, while in the rabbit, immune complexes are not involved.[67]

Washida[68] concluded that LPS is capable of binding to both platelets and granulocytes by means of an LPS cell membrane receptor.

The reticuloendothelial system (RES), although not an inhibitor of the LAL test, is capable of removing endotoxin from the systemic circulation. One means by which the RES achieves this reduction is by the activation of HDLs that have a clearing effect on both cholesterol and LPS as noted above.[69] The natural endpoint for cholesterol transport is the Kupffer cells in the liver. Carey et al.[70] demonstrated localization of labeled endotoxin within the liver after massive endotoxin infusion in experimental animals.

B. FALSE POSITIVES

False positive reactions in the LAL test have been reported in both the absence of Gram-negative and the presence of Gram-positive bacteremia.[71,72] Any compromise in the capacity of the liver — whether from this infection or liver dysfunction — to remove and assimilate endotoxin, will allow detection in the systemic circulation of rising levels derived from natural gut-bacterial sources.

Recent developments in the fractionation of the *Limulus* lysate have led to the recognition of two pathways that result in either a gel-clot or a positive chromogenic substrate reaction.[73] Pearson et al.,[42] in an investigation of hemodialyzer endotoxin LAL test reactive materials, reviewed a considerable number of substances. Many of these were high molecular weight substances including peptidoglycan, synthetic dextrans, carrageenan, β(1–3) D-glucan, mannans, and dextrans. All were subsequently discovered to react with the alternate pathway in the lysate test.[74] They further noted that these materials would not lead to pyrogenicity.

Many studies have shown little correlation of the LAL test with subsequent blood cultures for Gram-negative bacteremia.[75] In some instances, the sources of infection were found to be an abscess or other localized site of colonization. In the presence of normal hepatic function, a positive LAL test should lead to a search for a localized source of Gram-negative bacterial infection. A review of the common methods utilized for preparation of plasma samples for the LAL test follows.

C. HEATING AND DILUTION

Heat and dilution is the most commonly applied method for the removal of interfering substances from plasma samples in the quantitative LAL determination of endotoxin.[39,76,104] The use of 15 to 30 U of final concentration of heparin appears to increase the effectiveness of differentiating between those individuals having septic conditions and normal afebrile human subjects.[76,103,104]

Cooperstock et al.,[76] in a study of the role of endotoxin in Reye's syndrome, diluted plasma samples 1:3 in endotoxin-free water and then heated the samples at 100°C for 10 min. They found this method to be more reliable than the pH shift technique and easier to perform than chloroform extraction. Detection capability was as low as 0.1 ng/ml of *E. coli*, 026:B6 endotoxin.

Goldstein et al.[78] studied a variable dilution method of 1:10, 1:20, and 1:100 followed by heat at 95°C for 10 min. All dilutions were in endotoxin-free water. This approach was further evaluated by Nitsche et al.,[77] in plasma samples by both turbidimetric and chromogenic assays. They felt that the presence of fibrogen played a significant role, particularly in the turbidimetric assay, and noted a 1:40 dilution and heating to 80°C for 5 min was capable of making the fibrinogen contribution negligible in the chromogenic determination.

Dubose et al.[79] continued the heat and dilution technique of Cooperstock and reported endotoxin detection capability to 0.06 ng/ml. A switch from pyrogen-free water to 0.5 *M* phosphate buffer increased the sensitivity of the LAL test by a factor of 10. Heat and dilution was subsequently reported to negate the effects of both inhibitors and enhancers in plasma.[80-82]

Pearson et al.[39] conducted a study utilizing both the nephelometric and the chromogenic LAL tests. Their samples were diluted 1:10 with pyrogen-free water and heated at 70°C for 10 min. They measured endotoxin levels from 120 to 850 pg/ml in the blood of patients with culture-proven Gram-negative bacteremia.

Obayashi et al.[73] fractionated LAL into the endotoxin sensitive and glucan sensitive pathways and felt that in their system the heat and dilution method was unsatisfactory.

D. PLATELET EXTRACTION

The role of platelets in the clearance of endotoxin from blood by means of a binding reaction of LPS to platelets was demonstrated by Das et al.[83] Infusion of sub-lethal doses of *E. coli* endotoxin were readily cleared in 10 min, but when repeated in thrombocytopenic animals, clearance was retarded. Lysis of platelets resulted in the release of bound endotoxin and led to the author's suggestion that this procedure would be necessary to increase the sensitivity and reliability of the LAL test. Sturk et al.[84] tested the chromogenic assay in spiked samples *in vitro* and reported little relation to platelets, but noted that binding may be different in *in vivo* clinical situations.

In *in vitro* human studies,[83] 80% recovery was obtained when samples were spiked with 0.5 μg/ml endotoxin and found evenly distributed between the platelet layers and plasma (cell free). At lower concentrations (0.005 and 0.001 μg/ml) only 40% recovery was obtained. When fractions of ng/ml were utilized, endotoxin was detectable only in the platelet-rich fraction.

E. AMMONIUM SULFATE PRECIPITATION

Ammonium sulfate has been widely used as a method for fractionation of plasma proteins. Goto and Nakamura[85] reported a revised *Limulus* test using this method for the elimination of plasma protein inhibitors of gelation. They also reported their technique to be practical for endotoxin determination in saline, urine, transudate, exudate, CSF, and suspensions of *E. coli*. All results were read by interpretation of patterns on silicone coated glass slides. Their technique involved mixing 0.2 ml sample with 0.7 ml phosphate buffered saline (PBS) and 0.1 ml saturated ammonium sulfate. The mixture was heated to 100°C for 10 min, then centrifuged. From the supernatant, 0.2 ml was mixed with equal volumes of *Limulus* lysate (Pregel 1A, Teikoku Hormone Manufacturing Company, Japan) previously diluted 1:1 with PBS. The combination was incubated at 37°C for 30 min and the slide was evaluated. A homogeneous pattern was interpreted as a positive reaction. *E. coli* 0111:B4 LPS was detectable in physiologic saline at $^1/_5$ the detectable level (0.5 ng/ml) in plasma. No comparisons were noted with other existing methods. The authors felt this method exhibited a more objective endpoint.

F. ETHER METHOD

The precipitation of plasma proteins by ether for the elimination of inhibitors of the LAL test was evaluated by Harada et al.[86] Their group compared this method with acidification, heating and dilution, and chloroform extraction. Their results provided evidence that the ether method enhanced the sensitivity of the LAL test more than heating for 30 min at 56°C. They also reported the ether method to be more sensitive than the pH shift method.[87,88] Plasma at 1 ml was mixed with 0.25 ml of ethyl ether and shaken for 3 min. Ether was removed by vacuum treatment. Sensitivity was unchanged by either increasing the amount of ether or the shake time. A sensitivity of 10^{-1} μg/ml was achieved when the sample was heated at 56°C for 30 min, but when diluted 3-fold, followed by heating as above, the sensitivity increased to 10^{-4} μg/ml. When compared to a three-fold dilution followed by heating at 100°C for 10 min, a similar sensitivity of 10^{-4} μg/ml was obtained.[87] The chloroform method led to the same sensitivity but required a 4-h extraction rather than the 3 min required by the ether method. From these data, no advantage in terms of sensitivity could be ascertained over the heat and dilution method. One less step is required, and considerable time is saved. The authors also noted a tendency toward softening of gel endpoints with the ether method.

G. PERCHLORIC ACID

In a comparison of methods to eliminate inhibitors of the LAL test in a chromogenic procedure, Obayashi et al.[89] sought to eradicate nonendotoxin dependent amidase activities. They specifically attempted to remove activated Factor X and antithrombin III. Methods studied were perchloric acid (PCA), chloroform, heat and dilution, and pH shift. In this study, the PCA technique led to 100% recovery of *E. coli* 0111.B4 spiked samples in both plasma and serum. They further noted no effect by sample dilution or by various anticoagulants (heparin, citrate, or EDTA) utilized to collect blood. Their procedure involved the addition of 0.2 ml of 1.9% PCA to 0.1 ml platelet-rich plasma or serum followed by incubation at 37°C for 20 min. After centrifugation at 3000 RPM for 15 min, 0.1 ml of 0.2 sodium hydroxide was added to an equal volume of supernatant. Endotoxin was measured by means of a commercially prepared lyophilized amoebocyte lysate (*Tachypleus tridentatus*) and a synthetic chromogenic substrate ("Pyrodick":Seihagaku Hogyo LTD, Tokyo). Detection of endotoxin was less than 20 pg/ml in normal healthy subjects, while levels as high as 1000 pg/ml were noted among patients with sepsis, malignancy, leukemia, cirrhosis, and systemic lupus.

H. RESIN EXTRACTION

Affinity chromatography for the detection of endotoxin in 1.0 ml of heparinized plasma has been described.[90] This procedure was also applicable for detection of endotoxin in any body fluid. The authors prepared a polymeric matrix having endotoxin binding capacity. Endotoxin is "unmasked" by disassociation using salts, detergents, or organic chemicals. Their preference was 1.0 ml of 0.002 *M* benzamidine. Endotoxin is bound to the chromatography column, and contaminants washed by means of 2 ml of saline and 8 ml of 0.005 N acetic acid. Endotoxin is eluted with 5 ml of 2.0% sodium dodecyl sulfate, 3% sodium deoxycholate, or 2 *M* magnesium chloride. The eluate is precipitated with cold ethanol and incubated in 1 ml pyrogen-free water. Quantitation of endotoxin is performed with the LAL test. The authors also indicated that labeled antibody may be used to measure bound endotoxin on the column. No recoveries of spiked endotoxin were noted in this chromatographic procedure.

I. CHLOROFORM EXTRACTION

Removal of inhibitory factors by means of chloroform treatment of whole blood or plasma was reported by Levin et al.[46] in recovery studies with spiked endotoxin. This procedure utilized 1:4 chloroform extraction over a 1-h period followed by dilution and then tested by gel-clot. Although they reported nearly complete recovery, further investigations were able to demonstrate the efficacy of inhibitor removal by means of a 4-h chloroform extraction. The prolonged time required for chloroform extraction detracts from the efficiency of this procedure for the detection of endotoxin.

When endotoxin is added to a fixed amount of lysate, proportional amounts of protein are precipitated. Nandan et al.,[91] by means of this assayable precipitate, determined unknown endotoxin levels by comparison with a standard curve developed by means of *E. coli* endotoxin in known quantities. They felt this procedure to be more sensitive than the gelation method. Lowry protein precipitate was measured. Endotoxin could not be accurately determined in whole blood and platelet-rich plasma, but could be tested after platelet lysis. Inhibitory proteins were removed by chloroform extraction with vigorous mixing for 1h. *E. coli* endotoxin could be detected by this procedure to 0.195 ng/ml in whole blood and 0.049 ng/ml in platelet-rich plasma.

J. PH SHIFT METHOD

Adjustments of pH in plasma samples were begun on the assumption that endotoxin was bound to the globulin fraction and was released for assay by acid precipitation of the englobulin fraction.[87] Plasma pH was lowered to 4.0 with 25% acetic acid then returned to pH 6.2 by the addition of dibasic potassium phosphate buffer. The authors reported a sensitivity of measurement from 1 to 5 ng/ml of plasma. The use of the pH shift method and its applicability to clinical evaluation were confirmed by Jones and Roe in weekly endotoxin measurements of the plasma of patients in a burn unit.[88]

VI. CURRENT STATUS OF THE LAL TEST

The LAL test has developed into a sensitive and specific test for the quantitative measurement of bacterial endotoxin. Although the most difficult fluid to analyze for the presence of endotoxin is blood, the method most frequently used to remove extraneous factors in the LAL test has been the heat and dilution technique. Dilution has varied from 1:3 with heating to 100°C for 10 min, to 1:10 dilution at 75°C for 10 min.[41]

In the evolution of man, selection would have undoubtedly promoted those individuals who exhibit endotoxin-neutralizing factors in their plasma. These factors, including the human coagulation system (noted to be similar to the clotting mechanism of the horseshoe crab),[92] would naturally interfere in the sensitivity of the LAL test to endotoxin. Differences have been noted even between plasma and serum,[93] and thought to be due to an acute phase factor. du Moulin et al.,[93] in a study of the detection of Gram-negative bacteria in the blood of rats, found an insensitivity to the presence of *Bacteroides fragilis*, which they noted lacks the lipid A component.

Buller et al.,[95] in a prospective study of 56 postoperative patients, correlated endotoxemia and bacteremia. They reported a sensitivity of 92%, and a specificity of 91% with a positive predictive value of 75% and a negative predictive value of 98%. Although some blood cultures revealed no growth while endotoxin was noted to be present, many of their patients were receiving broad-spectrum antibiotics. It is significant that their chromogenic tests for endotoxin were positive shortly before (12 to 24 h) the onset of clinical symptoms.

Two activation pathways in the LAL test have been found.[73] One begins with endotoxin sensitive factor C and the other with glucan sensitive factor G (see Chapter 3). The factor G pathway is activated by many of the materials found in use in hemodialyzer systems. Fungal polysaccharides also interact with this pathway. Pearson[74] recently covered the relationship of false-positive LAL-reactive material in the LAL test system and false-positive reactions to commercial products. Only endotoxin is capable of reacting with the factor C pathway.

Dolan et al.[96] evaluated 520 septic episodes in a hospital setting and found a sensitivity of 21% and a specificity of 93%. Their predictive value for a positive test was 79%. They further attempted repeat assays when the LAL test was positive but blood cultures were negative. In two cases, occult Gram-negative intraabdominal abscesses were subsequently discovered.

Pearson et al.[39] evaluated the LAL chromogenic and nephelometric assay systems with 1:10 dilution followed by heating at 70°C for 10 min in Gram-negative bacterial sepsis. Their study of plasma collected from 45 patients found a strong association between a positive LAL test and a septic condition. They noted a sensitivity and specificity of 90 and 94%, respectively.

In another study, endotoxin levels in peripheral blood were correlated with Gram-negative peritonitis.[76] They found a wide range of Gram-negative bacteria including *B. fragilis* and indicated a good correlation between endotoxin levels and infection, but did not report specificity or sensitivity levels. They did note that surgical outcome and recovery bore a direct relationship to rising or falling endotoxin levels.

McCartney and co-workers,[97] in a prospective study using the chromogenic LAL test, found little correlation with blood cultures. This finding lends credence to the utility of the LAL test in those situations wherein the Gram-negative infection is not related to generalized endotoxemia. It would appear that endotoxin is readily capable of presentation into the systemic circulation regardless of the source or localization of the infection.

Zhang et al.[98] combined the LAL test with the enzyme-linked immunosorbent assay (ELISA)

TABLE 1
Sensitivity of Four Assay Systems in the Detection of Endotoxin

Method	Sensitivity (pg/ml)
Rabbit pyrogen	100
LAL gel-clot	10
LAL chromogenic	1
LAL ELISA	0.1

Modified from Friberger, P., *Am. Clin. Prod. Rev.*, 6(4), 12, 1987.

directed by a monoclonal antibody to purified coagulogen. The test system is based on the observation that loss of coagulogen by the production of clot-gel from the effects of endotoxin results in the loss of antigenicity to coagulogen. They had previously reported the development of a quantitative LAL rocket immunoelectrophoretic method that was sensitive and reproducible but required highly skilled operators. The LAL-ELISA reportedly has a sensitivity of 0.1 pg control standard endotoxin (CSE), and required only 1/160 of the amount of LAL used in the gel-clot method. The LAL–ELISA system requires 0.01 ml of sample in comparison to 0.1 ml for most other LAL tests. Blood samples were treated by 10-fold dilution with subsequent heating at 75°C for 10 min. Recovery of CSE (2 ng/ml) was reportedly 85.3%. The secondary (1–3)-β-D-glucan pathway was less pronounced than in standard chromogenic assays and could easily be differentiated from endotoxin by kinetic differences. (1–3)-β-D-Glucan was selectively examined and found to be 1000 times less reactive than endotoxin and evidenced a much slower reaction with LAL.

Comparisons of the sensitivity of various methods used for detection of endotoxin reveal consistent improvement in detection capability (Table 1).[99] The LAL-ELISA may be the most significant step forward in endotoxin determination since the addition of a chromogenic substrate. The test affords easy performance, reproducibility, high sensitivity, and a lower interference from plasma factors than existing systems. Further refinements should make the LAL-ELISA a valuable tool for the quantitation of endotoxin in clinical specimens.

VII. FUTURE CLINICAL TREND

While earlier studies designed to evaluate the presence of circulating endotoxin in the bloodstream were relatively qualitative, more recent studies have focused on quantitative measurements using the chromogenic substrate assay.[100] In this study, 95% (45 episodes) of the afebrile patients had endotoxin levels of less than 5 pg/ml while only 3% had between 5 and 10 pg/ml and no individuals had circulating levels in excess of 10 pg/ml. In contrast, 31% of a septic population had endotoxin levels in excess of 10 pg/ml. However, 58% of this population had levels less than 5 pg/ml while 11% had between 5 and 10 pg/ml. These observations suggest that individuals having more than 10 pg/ml endotoxin have a significant probability of being septic.

These observations were confirmed and extended in another study which evaluated plasma samples from normal afebrile human subjects.[39] In 120 evaluations, the background endotoxin on average was less than 1.0 pg/ml while individual samples ranged from 0.0 to 16.6 pg/ml. In contrast, 11 of 12 patients diagnosed as having Gram-negative bacteremia exhibited endotoxin levels between 118 and 850 pg/ml referenced to *E. coli* 055:B5 endotoxin. Five individuals having Gram-positive bacteremia were negative for plasma endotoxin. While these levels are consistently higher than in the study reported by van Deventer and co-workers, these data suggest that a clear demarcation can be made consistently between normal human subjects and patients with Gram-negative sepsis.

As was discussed, sample preparation is critical to obviate false positives and negatives when evaluating plasma for endotoxin concentration. It is interesting that the two studies just described were able to discriminate between septic and nonseptic individuals and each study reported similar sample preparation. In both cases whole blood was collected in 15 to 30 U final concentration of nonpyrogenic heparin. Platelet rich plasma (PRP) samples were then diluted 1:10 and subsequently heated at 70°C for 10 min or 75°C for 15 min.[101] Since these methods are remarkably similar in sample preparation, it is tempting to speculate that this approach or one very close to it provides the foundation for an LAL test that can consistently identify those individuals who have a septic condition. Subsequently, PRP was prepared by centrifugation of whole blood at 150 or 100 × *g* for 10 min.

Although plasma levels of endotoxin have been reported to be high in neutropenic patients undergoing cytoreductive therapy, there was no correlation between level of endotoxin and degree of fever. In this study, normal controls showed mean endotoxin levels of 3 pg/ml compared to 4.6 and 6.0 pg/ml in febrile and afebrile patients, respectively.[102] In a separate study, 52% of 21 episodes of fever in patients with a hematological malignancy who were infected revealed evaluated endotoxin (as measured in optical density units).[103] Endotoxemia was demonstratable in 27% of 22 episodes of fever of unknown origin. In 45 afebrile patients neither neutropenia nor cytotoxic chemotherapy was the cause of circulating endotoxin. The author concluded that the passage of endotoxin from portal blood into the systemic circulation can contribute to unexplained fever in immunocompromised patients.[103]

The LAL test is the most sensitive and specific measure of endotoxin from Gram-negative bacteria.[104] Refinements in the Pathway C LAL test of Obayashi[73] may continue the increase in specificity that has developed from Levin and Bang's original method to the modification using a chromogenic substrate.[105]

While the LAL test may be specific for endotoxin, bacterial cultures must still be performed for the identification of the specific microorganism. The major use of the LAL test may be in the clinical setting with identification that in progress is a process that has resulted in endotoxin presence in the systemic circulation. In this situation, a therapeutic monoclonal antibody may be used in the future to neutralize tumor necrosis factor or the active core site of the lipopolysaccharide.[61] In order to determine the specific etiology of the source of endotoxin, all clinical parameters, including intestinal leak syndromes, pharmaceutical product contamination, functional capacity of the liver and source(s) of Gram-negative infection or colonization will require evaluation. Although early intervention may necessitate the treatment of a few false positives, the return in terms of preventing endotoxin-induced shock may make this approach worthwhile.

Other parameters must be entertained before the questions surrounding the etiology of shock are resolved. IL-1 was recently reported to be capable of inducing a shock-like state in rabbits, and may have a synergistic relationship to tumor necrosis factor as well as yet unknown interrelationships with other cytokines.[106]

REFERENCES

1. **Aune, S., Aasen, A. O., Kierulf, P., and Stromme, J. H.,** Quantitative determination of endotoxin in plasma by means of a new chromogenic peptide substrate. A methodological study, *Acta. Chir. Scand. Suppl.*, 509, 1982.
2. **Gaffin, S. L.,** Control of septic shock — present day concept, *S. A. J. Hosp. Med.*, January, 4, 1982.
3. **Shenep, J. L., Flynn, P. M., Barrett, F. F., Stidham, G. L., and Westenkirchner, D. F.,** Serial quantitation of endotoxemia and bacteremia during therapy for Gram-negative bacterial sepsis, *J. Infect. Dis.*, 157, 565, 1988.
4. **Kelsey, M. C., Lipscomb, A. P., and Mowles, J. M.,** Limulus amoebocyte lysate endotoxin test: an aid to the diagnosis in the septic neonate, *J. Infect.*, 4, 69, 1982.

5. **Cooperstock, M. and Riegle, L.,** Plasma *Limulus* gelation assay in infants and children: correlation with Gram-negative bacterial infection and evidence for "intestinal endotoxemia", in *Bacterial Endotoxins: Structure, Biomedical Significance, and Detection with the Limulus Amebocyte Lysate Test*, Ten Cate, J. W., Buller, H. R., and Sturk, A., Eds., Alan R. Liss, New York, 1985, 239.
6. **Karakusis, P. H.,** Considerations in the therapy of septic shock, *Med. Clin. North Am.*, 70, 993, 1986.
7. **Morrison, D. C. and Ryan, J. C.,** Endotoxin and disease mechanisms, *Annu. Rev. Med.*, 38, 417, 1987.
8. **Dinarello, C.,** Interleukin-1, *Rev. Infect. Dis.*, 6, 51, 1984.
9. **Beutler, B. and Cerami, A.,** Cachectin: more than a tumor necrosis factor, *N. Engl. J. Med.*, 316, 379, 1987.
10. **Vukajlovitch, S. W., Hoffman, J., and Morrison, D. C.,** Activation of human serum complement by bacterial lipopolysaccharides: structural requirements for antibody independent activation of the classical and alternative pathways, *Mol. Immunol.*, 24, 319, 1987.
11. **Robinson, J. A., Kondnycky, M. L., Loeb, H. S., Racic, M. R., and Gunnar, R. M.,** Endotoxin prekallikrein, complement and systemic vascular resistance — sequential measurements in man, *Am. J. Med.*, 59, 61, 1975.
12. **Levy, G. A., Schwartz, B. S., Curtiss, L. K., and Edginton, T. S.,** Regulatory roles of Tu and T cells in the collaborative cellular initiation of the extrinsic coagulation pathway by bacterial lipopolysaccharide, *J. Clin. Invest.*, 76, 548, 1985.
13. **Friberger, P.,** A new method of endotoxin determination, *Am. Clin. Prod. Rev.*, 6, 12, 1987.
14. **Michalek, S. M., Moore, R. N., McGhee, J. R., Rosenstreich, D. L., and Mergenhagen, S. E.,** The primary role of lymphoreticular cells in the mediation of host responses to bacterial endotoxin, *J. Infect. Dis.*, 41, 55, 1980.
15. **Dinarello, C. A. and Mier, J.,** Lymphokines, *N. Engl. J. Med.*, 317, 940, 1987.
16. **Beutler, B. and Cerami, A.,** Cachectin and tumor necrosis factor as two sides of the same biological coin, *Nature*, 320, 582, 1986.
17. **Beutler, B., Milsark, J. W., and Cerami, A.,** Cachectin — tumor necrosis factor: production, distribution, and metabolic fate *in vivo*, *J. Immunol.*, 135, 3972, 1985.
18. **Michie, H. R., Manogue, K. R., Spriggs, D. R., Ravhaug, A., O'Dwyer, S., Dinarello, C. A., Wolff, S. M., and Wilmore, D. W.,** Detection of circulating TNF after endotoxin administration, *N. Engl. J. Med.*, 318, 1481, 1988.
19. **Dinarello, C. A., Cannon, J. G., Wolff, S. M., Bernheim, H. A., Beutler, B., Cerami, A., Figari, I. S., Palladino, M. A., Jr., and O'Connor, J. V.,** Tumor necrosis factor (cachectin) is an endogenous pyrogen and induces production of interleukin-1, *J. Exp. Med.*, 163, 1433, 1986.
20. **Cooper, E. S., Flaum, M. A., and Springgate, C. F.,** Macrophages: controllers of the human immune system, *Mod. Med.*, 55, 40, 1987.
21. **Beutler, B., Milsark, J. W., and Cerami, A. C.,** Passive immunization against cachectin/tumor necrosis factor protects mice from lethal effect of endotoxin, *Science*, 229, 869, 1985.
22. **Teng, N. N. H., Kaplan, H. S., Hebert, J. M., Moore, C., Douglas, H., Werderlich, A., and Braude, A. I.,** Protection against gram negative bacteremia and endotoxemia with human monoclonal IgM antibodies, *Proc. Natl. Acad. Sci. U.S.A.*, 82, 1790, 1985.
23. **Young, L. S.,** Functional activity of monoclonal antibodies against lipopolysaccharide (LPS) antigens of gram negative bacilli, *Clin. Res.*, 32, 518A, 1984.
24. **Ziegler, E. J., Teng, N. N. H., Douglas, H. Wunderlich, A., Berger, H. J., and Balmer, S. D.,** Treatment of *Pseudomonas* bacteremia in neutropenic rabbits with human monoclonal IgM antibody against *E. coli* lipid A, *Clin. Res.*, 35, 619A, 1987.
25. **Mathison, J. C., Wolfson, E., and Ulevitch, R. J.,** Participation of tumor necrosis factor in the mediation of gram negative bacterial lipopolysaccharide induced injury to rabbits, *J. Clin. Invest.*, 81, 1925, 1988.
26. **Sieff, C. A., Niemeyer, C. M., Mentzer, S. J., and Faller, D. V.,** Interleukin-1, tumor necrosis factor, and the production of colony stimulating factors by cultured mesenchymal cells, *Blood*, 72, 1316, 1988.
27. **Sheagren, J. N.,** Septic shock and corticosteroids, *N. Engl. J. Med.*, 305, 456, 1981.
28. **Jacob, H. S., Craddock, P. R., Hammerschmidt, D. E., and Moldow, C. F.,** Complement-induced granulocyte aggregation: an unsuspected mechanism of disease, *N. Engl. J. Med.*, 302, 789, 1980.
29. **Mathison, J. C. and Ulevitch, R. J.,** *In vivo* interaction of bacterial lipopolysaccharide (LPS) with rabbit platelets: modulation by C3 and high density lipoproteins, *J. Immunol.*, 126, 1575, 1981.
30. **Gaffin, S. L.,** Control of septic shock — present day concept, *S. A. J. Hosp. Med.*, January, 4, 1982.
31. **Morrison, D. C. and Cochrane, C. G.,** Direct evidence for Hageman factor (factor XLL) activation by bacterial lipopolysaccharides (endotoxins), *J. Exp. Med.*, 140, 797, 1974.
32. **Aulitzky, W. E., Tilg, H., Herold, M., Berger, M., Vogel, W., Judmaier, G., Gastl, G., Mull, B., Flener, R., Wiegele, J., Pichler, E., Denz, H., Goheim, E., Aulitzky, W. K., and Huber, C.,** Enhanced serum levels of B-2 microglobulin, neopterin, and interferon-gamma in patients treated with recombinant tumor necrosis factor-alpha, *J. Interferon Res.*, 8, 655, 1988.
33. **Harris, R. L., Musher, D. M., Bloom, K., Gathe, J., Rice, L., Sugarman, B., Williams, T. W., and Young, E. J.,** Manifestations of sepsis, *Arch. Intern. Med.*, 147, 1895, 1987.

34. **Clark, S. C. and Kamen, R.,** The human hematopoietic colony-stimulating factors, *Science*, 236, 1229, 1987.
35. **Valmari, P. and Peltola, H.,** Serum C-reactive protein: a valuable differentiator between viral and bacterial meningitis, *Infect. Med.*, September, 308, 1987.
36. **Levin, J. and Bang, F. B.,** The role of endotoxin in the extracellular coagulation of *Limulus* blood, *Bull. Johns Hopkins Hosp.*, 115, 265, 1964.
37. **Zhang, G., Baek, L., and Koch, C.,** New microassay for quantitation of endotoxin using *Limulus* amebocyte lysate enzyme-linked immunosorbent assay, *J. Clin. Microbiol.*, 26, 1464, 1988.
38. **Elin, R.,** Clinical utility of the Limulus test with blood, CSF, and synovial fluid, in *Biomedical Applications of the Horseshoe Crab (Limulidae)*, Bang, F. B., Levin, J., Marchalonis, J. J., Pistole, T. G., Prendergast, R. A., Shuster, C., and Watson, S. W., Eds., Alan R. Liss, New York, 1979, 279.
39. **Pearson, F. C., Dubczak, J., Weary, M., Bruszer, G., and Donohue, G.,** Detection of endotoxin in plasma of patients with gram negative bacterial sepsis by the *Limulus* amoebocyte lysate assay, *J. Clin. Microbiol.*, 21, 865, 1985.
40. **Ditter, B., Becker, K. P., Urbaschek, R., and Urbaschek, B.,** Detection of endotoxin in blood and other specimens by evaluation of photometrically registered LAL-reaction-kinetics in microtiter plates, in *Endotoxins and Their Detection with the Limulus Amebocyte Lysate Test*, Watson, S. W., Levin, J., and Novitsky, T. J., Eds., Alan R. Liss, New York, 1982, 387.
41. **Elin, R. J.,** Clinical utility of the *Limulus* amebocyte lysate (LAL) test, in *Bacterial Endotoxins: Structure, Biomedical Significance, and Detection with the Limulus Amebocyte Lysate Test*, Ten Cate, J. W., Buller, H. R., and Sturk, A., Eds., Alan R. Liss, New York, 1985, 307.
42. **Pearson, F. C., Dubczak, J., Weary, M., and Anderson, J.,** Determination of endotoxin levels and their impact on interleukin-1 generation in continuous ambulatory peritoneal dialysis and hemodialysis, *Blood Purif.*, 6, 207, 1988.
43. **Pearson, F. C., Bohon, J., Lee, W., Bruszer, G., Sagona, M., Jakubowski, G., Dawe, R., Morrison, D., and Dinarello, C.,** Characterization of *Limulus* amoebocyte lysate-reactive material from hollow-fiber dialyzers, *Appl. Environ. Microbiol.*, 48, 1189, 1984.
44. **Pearson, F. C.,** Pyrogens: endotoxins, LAL testing, and depyrogenation, in *Advances in Parental Sciences*, Vol. 2, Robinson, J., Ed., Marcel Dekker, New York, 1985, 232.
45. **Oroszlan, S., McFarland, V., Mora, P. T., and Shear, M. J.,** Reversible inactivation of an endotoxin by plasma proteins, *Ann. N. Y. Acad. Sci.*, 133, 622, 1966.
46. **Levin, J., Tomasulo, P. A., and Oser, R.,** Detection of endotoxin in human blood and demonstration of an inhibitor, *J. Lab. Clin. Med.*, 75, 903, 1970.
47. **Johnson, K. J. and Ward, P. A.,** The requirement for serum complement in the detoxification of bacterial endotoxin, *J. Immunol.*, 108, 611, 1972.
48. **Johnson, K. J., Ward, P. A., Goralnick, S., and Osborn, M. J.,** Isolation from human serum of an inactivator of bacterial lipopolysaccharide, *Am. J. Pathol.*, 88, 559, 1977.
49. **Skarnes, R. C. and Rosen, F. S.,** Host-dependent detoxification of bacterial endotoxin, in *Microbiol Toxins: A Comprehensive Treatise*, Vol. 5, Ajl, S. J., Kadis, S., and Weinbaum, G., Eds., Academic Press, New York, 1971, 151.
50. **Skarnes, R. C.,** The inactivation of endotoxin after interaction with certain proteins or normal serum, *Ann. N. Y. Acad. Sci.*, 133, 644, 1966.
51. **Skarnes, R. C.,** Host defense against bacterial endotoxemia: mechanism in normal animals, *J. Exp. Med.*, 132, 300, 1970.
52. **Rosen, F. S., Skarnes, R. C., and Landy, M.,** Inaction of endotoxin by humoral component. III. Role of divalent cation and dialyzable component, *J. Exp. Med.*, 108, 701, 1958.
53. **Keene, W. R., Landy, M., and Shear, M.,** Inactivation of endotoxin by humoral component. VII. Enzymatic degradation of endotoxin by blood plasma, *J. Clin. Invest.*, 40, 302, 1961.
54. **Von Eschen, K. B. and Rudbach, J. A.,** Inactivation of endotoxin by serum: a phylogenetic study, *J. Infect. Dis.*, 129, 21, 1974.
55. **Nakamura, S. S. and Iwanaga, S.,** Haemostasis and coagulation in horseshoe crab, *Tanpakushitsu Kakusan Koso* (Protein, Nucleic Acid and Enzyme), 23, 277, 1978.
56. **Young, N. S., Levin, J., and Prendergast, R.,** An invertebrate coagulation system activated by endotoxin: evidence for enzymatic medication, *J. Clin. Invest.*, 51, 1790, 1972.
57. **Kobayashi, M. and Yamamoto, M.,** Studies on the gelation reaction of *Limulus* lysate (pre-gel). V. Purification of the inhibitory substances of the gelation reaction of *Limulus* lysate from human plasma, *Yakugaku Zasshi*, 100, 500, 1980.
58. **Gruning, W.,** Fur die Gesamte Physiologie des Menschen und der Tieren, *Pflugers Arch.*, 247, 292, 1943.
59. **Peanosky, R. J. and Laskowski, M.,** Partial purification of the trypsin inhibitor from blood plasma, *J. Biol. Chem.*, 204, 153, 1953.
60. **Koster, F., Levin, J., Walker, L., Tung, K., Gilman, R. H., Rahaman, M. M., Majid, M. A., Islam, S., and Williams, R. C.,** Hemolytic-uremic syndrome after shigellosis, *N. Engl. J. Med.*, 298, 927, 1978.

61. **Ziegler, E. J.,** Perspective: protective antibody to endotoxin core: the emperor's new clothes?, *J. Infect. Dis.*, 158, 286, 1988.
62. **Ulevitch, R. J., Johnston, A. R., and Weinstein, D. B.,** New function for high density lipoproteins, *J. Clin. Invest.*, 64, 1516, 1979.
63. **Moreau, S. C. and Skarnes, R. C.,** Host resistance to bacterial endotoxemia mechanisms in endotoxin tolerant animals, *J. Infect. Dis. Suppl.* 128, 122, 1973.
64. **Braude, A. I., Carey, F. J., and Zalesky, M.,** Studies with radioactive endotoxin. II. Correlation of physiologic effects with distribution of radioactivity in rabbits injected with lethal dose of *E. coli* endotoxin labelled with radioactive sodium chromate, *J. Clin. Invest.*, 34, 858, 1955.
65. **Hawiger, J., Hawiger, A., and Timmons, S.,** Endotoxin sensitive membrane component of human platelets, *Nature*, 256, 125, 1975.
66. **Thomas, D. P., Wessler, S., and Reimer, S. M.,** *Fed. Proc.*, 21, 61, 1962.
67. **Myllyla, G.,** Aggregation of human blood platelets by immune complexes in the sedimentation pattern test, *Scand. J. Haematol. Suppl.* 19, 1, 1973.
68. **Washida, S.,** Endotoxin receptor site. II. Specificity of endotoxin receptor of platelets and sensitivity to endotoxin *in vivo*, *Acta Med. Okayama*, 32, 217, 1978.
69. **Nimer, S. D., Champlin, R. E., and Golde, D. W.,** Serum cholesterol-lowering activity of granulocyte-macrophage colony stimulating factor, *JAMA*, 260, 3297, 1988.
70. **Carey, F. J., Braude, A. I., and Zalesky, M.,** Studies with radioactive endotoxin. III. The effect of tolerance on the distribution of radioactivity after intravenous injection of *Escherichia coli* endotoxin labeled with Cr51, *J. Clin. Invest.*, 37, 441, 1958.
71. **Levin, J., Poore, T. E., Young, M. S., Margolis, S., Zauker, N. P., Townes, A. S., and Bell, W. R.,** Gram-negative sepsis: detection of endotoxemia with the *Limulus* test, *Ann. Intern. Med.*, 1, 76, 1972.
72. **Stumacher, R. J., Kovnat, M. J., and McCabe, W. R.,** Limitations of the usefulness of the *Limulus* assay for endotoxin, *N. Engl. J. Med.*, 288, 1261, 1973.
73. **Obayashi, T., Tamura, H., Tanaka, S., Ohki, M., Takahashi, S., Arai, M., Masuda, M., and Kawai, T.,** Removal of *Limulus* test-interfering factors in blood samples with perchloric acid and the improvement of the specificity of the *Limulus* test by fractionating amebocyte lysate, in *Detection of Bacterial Endotoxins with the Limulus Amebocyte Lysate Test*, Watson, S. W., Levin, J., and Novitsky, T. J., Eds., Alan R. Liss, New York, 1987, 369.
74. **Pearson, F. C.,** Relative importance of membrane *Limulus* amebocyte lysate-reactive material and ethylene oxide in hypersensitivity reactions, *Contr. Nephrol.*, 59, 134, 1987.
75. **Van Deventer, S. J. H., Buller, H. R., Ten Cate, J. W., Sturk, A., and Pau, W.,** Endotoxemia: an early predictor of septicaemia in febrile patients, *Lancet*, 1, 605, 1988.
76. **Cooperstock, M. S., Tucker, R. P., and Baublio, J. V.,** Possible pathogenetic role of endotoxin in Reye's syndrome, *Lancet*, 1, 1272, 1975.
77. **Nitsche, D., Kriewitz, M., Seifert, J., and Hamelmann, H.,** Investigations on an essential plasma factor disturbing the photometric determination of endotoxin in plasma samples with the *Limulus* test, in *Detection of Bacterial Endotoxins with the Limulus Amebocyte Lysate Test*, Watson, S. W., Levin, J., and Novitsky, T. J., Eds., Alan R. Liss, New York, 1987, 331.
78. **Goldstein, J. A., Reller, L. B., and Long, W. L.,** *Limulus* amebocyte lysate test in neonates, *Am. J. Clin. Pathol.*, 66, 1012, 1978.
79. **Dubose, D. A., Lemaire, M., Bosamaia, K., and Rolands, J.,** Comparison of plasma extraction techniques in preparation of samples for endotoxin testing by the *Limulus* amebocyte lysate test, *J. Clin. Microbiol.*, 11, 68, 1980.
80. **Thomas, L. L., Sturk, M. A., Kahle, L. H., and Ten Cate, J. W.,** Quantitative endotoxin determination in blood with a chromogenic substrate, *Clin. Chim. Acta*, 115, 63, 1981.
81. **Webster, J. C.,** Principles of a quantitative assay for bacterial endotoxins in blood that uses *Limulus* lysate and a chromogenic substrate, *J. Clin. Microbiol.*, 12, 644, 1980.
82. **Oune, S., Aasen, A. O., Kierulf, P., and Stromme, J. H.,** Quantitative determination of endotoxin in plasma by means of a new chromogenic peptide substrate. A methodological study, *Acta Chir. Scand. Suppl.* 509, 55, 1982.
83. **Das, J., Schwartz, A. A., and Folkman, J.,** Clearance of endotoxin by platelets: role in increasing the accuracy of *Limulus* gelation test and in combating experimental endotoxemia, *Surgery*, 74, 235, 1973.
84. **Sturk, A., Janssen, M. E., Muylaert, F. R., Joop, K., Thomas, L. L. M., and Ten Cate, J. W.,** Endotoxin testing in blood, in *Detection of Bacterial Endotoxins with the Limulus Amebocyte Lysate Test*, Watson, S. W., Levin. J., and Novitsky, T. J., Eds., Alan R. Liss, New York, 1987, 371.
85. **Goto, H. and Nakamura, S.,** Dry up method as a revised *Limulus* test with a new technique for gelation inhibitor removing, *Jpn J. Exp. Med.*, 49, 19, 1979.
86. **Harada, T., Nakahara, C., and Niwa, M.,** Studies on the elimination techniques of plasma inhibitor in the *Limulus* test, *Kasaoka Symp. Horseshoe Crabs Contrib. Med. Sci.*, 55, 1976.

87. **Reinhold, R. B. and Fine, J.,** A technique for quantitative measurement of endotoxin in human plasma, *Proc. Soc. Exp. Biol. Med.*, 137, 334, 1971.
88. **Jones, R. J. and Roe, E. A.,** Measurement of endotoxins with the *Limulus* test in burned patients, *J. Hyg. (Cambridge)*, 83, 151, 1979.
89. **Obayoshi, T., Kawai, T., Tomura, H., and Nakahara, C.,** New *Limulus* amebocyte lysate test for endotoxemia, *Lancet*, i, 289, 1982.
90. **Firca, J. R. and Rudbach, J. A.,** Method for systematic endotoxin detection, U.S. Patent 4,276,050, June 30, 1981.
91. **Nandan, R., Nakashima, C. T., and Brown, D.,** Detection of endotoxins in human blood and plasma. An improved *in vitro* pyrogen test, *Clin. Chem.*, 23, 2080, 1977.
92. **Juhlin, L. and Shelley, W. B.,** Radially oriented fabrin crystallization, a new test for endotoxin sensitivity in man, *Lancet*, 1, 21, 1977.
93. **du Moulin, G. C., Lynch, S. E., Hedley-Whyte, J., and Broitman, S. A.,** The detection of Gram-negative bacteria in the blood of septic rats by a *Limulus* amebocyte lysate assay, in *Detection of Bacterial Endotoxins with the Limulus Amebocyte Lysate Test*, Watson, S. W., Levin, J., and Novitsky, T. J., Eds., Alan R. Liss, New York, 1987, 349.
94. **Warren, H. S., Novitsky, T. J., Martin, P., Roslansky, P. F., and Siber, G. R.,** Endotoxin neutralizing capacity of sera from different patient populations assessed by the *Limulus* lysate test, in *Detection of Bacterial Endotoxins with the Limulus Amebocyte Lysate Test*, Watson, S. W., Levin, J., and Novitsky, T. J., Eds., Alan R. Liss, New York, 1987, 341.
95. **Buller, H. R., Ten Cate, J. W., Sturk, A., and Thomas, L. L.,** Validity of the endotoxin assay in post-surgical patients, in *Bacterial Endotoxins: Biomedical Significance, and Detection with the Limulus Amebocyte Lysate Test*, Ten Cate, J. W., Buller, H. R., Sturk, A., and Levin, J., Eds., Alan R. Liss, New York, 1985, 405.
96. **Dolan, S. A., Riegle, L., Berzofsky, R., and Cooperstock, M.,** Clinical evaluation of the plasma chromogenic *Limulus* assay, in *Detection of Bacterial Endotoxins with the Limulus Amebocyte Lysate Test*, Watson, S. W., Levin, J., and Novitsky, T. J., Eds., Alan R. Liss, New York, 1987, 405.
97. **McCartney, A. C., Piotrowicz, B. I., Edlin, S. E., and Ledingham, I. McA.,** Evaluation of the chromogenic *Limulus* lysate assay in septic shock, in *Detection of Bacterial Endotoxins with the Limulus Amebocyte Lysate Test*, Watson, S. W., Levin, J., and Novitsky, T. J., Eds., Alan R. Liss, New York, 1987, 459.
98. **Zhang, G., Baek, L., and Koch, C.,** New microassay for quantitation of endotoxin using *Limulus* amebocyte lysate combined with enzyme-linked immunosorbent assay, *J. Clin. Microbiol.*, 26, 1464, 1988.
99. **Friberger, P.,** A new method of endotoxin determination, *Am. Clin. Prod. Rev.*, 6, 12, 1987.
100. **van Deventer, S. J. H, Pauw, W., Ten Cate, J. W., Janssen, M. E., Buller, H. R., and Sturk, A.,** Clinical evaluation in febrile patients of an optimized endotoxin assay in blood, in *Detection of Bacterial Endotoxins with the Limulus Amebocyte Lysate Test*, Watson, S. W., Levin, J., and Novitsky, T. J., Eds., Alan R. Liss, New York, 1987, 489.
101. **Sturk, M. E., Janssen, F. R., Muylaert, J. K., Thomas, L. L. M., and Ten Cate, J. W.,** Endotoxin testing in blood, in *Detection of Bacterial Endotoxins with the Limulus Amebocyte Lysate Test*, Watson, S. W., Devin, J., and Novitsky, T. J. Eds., Alan R. Liss, New York, 1987, 371.
102. **Shands, W. J. and McKimmey, C.,** Plasma endotoxin increased levels in neutropenic patients do not correlate with fever, *J. Infect. Dis.*, 159(4), 777, 1989.
103. **Harris, R. I., Stone, P. C. W., Evans, G. R., and Stuart, J.,** Endotoxemia as a cause of fever in immunosuppressed patients, *J. Clin. Pathol.*, 37, 467, 1984.
104. **Pearson, F. C., Caruana, R., Burkart, J., Katz, D. V., Chenoweth, D., Dubczak, J., Bohon, J., and Weary, M.,** The use of the *Limulus* lysate assay to monitor hemodialyzer-associated soluble cellulosic material (LAL-reactive material), in *Detection of Bacterial Endotoxins with the Limulus Amebocyte Lysate Test*, Watson, S. W., Levin, J., and Novitsky, T. J., Eds., Alan R. Liss, New York, 1987, 211.
105. **Scully, M. F., Newman, Y. M., Clark, S. E., and Kakkar, V. V.,** Evaluation of a chromogenic method for endotoxin measurement, *Thromb. Res.*, 20, 263, 1980.
106. **Ziegler, E. J.,** Tumor necrosis factor in humans, *N. Engl. J. Med.*, 318, 1533, 1988.

Chapter 6

DETECTION OF GRAM-NEGATIVE BACTERIAL MENINGITIS

Ronald Nachum

TABLE OF CONTENTS

I. Introduction 68

II. Laboratory Diagnosis of Bacterial Meningitis 68

III. LAL Test for Gram-Negative Meningitis 70
A. Review of Literature 70
B. Efficacy of LAL Screening for Gram-Negative Bacterial Meningitis 74

IV. Conclusions and Future Trends 77

References 78

I. INTRODUCTION

Bacterial meningitis is a life-threatening infection requiring rapid clinical and laboratory diagnosis so that immediate and appropriate antimicrobial therapy can be administered to prevent brain damage and death. Common signs and symptoms of bacterial (purulent) meningitis include fever, headache, lethargy, vomiting, and meningeal irritation as manifested by demonstration of Kernig's and Brudzinski's signs.[1] Cutaneous lesions are of diagnostic importance in children with sepsis and meningitis. Purpura and petechiae are commonly observed with meningococcal disease and in patients with meningitis caused by other bacterial pathogens.[2] Seizures may occur in about 20 to 30% of patients, and various focal neurological signs may be present.[3] For a complete discussion of signs, symptoms and clinical diagnosis of bacterial meningitis, see References 2 and 3.

Bacterial meningitis is a major cause of morbidity and mortality in the U.S., with some 20,000 cases occurring on a yearly basis.[4] In 1978 the reported annual incidence was 2.9 cases per 100,000 population,[5] a figure estimated to be low due to underreporting.[6] Acute bacterial meningitis occurs predominantly in children under the age of five, with incidence and etiologic agent age associated. Neonatal disease is caused primarily by organisms indigenous to the maternal genital tract,[7] including *Escherichia coli*, group B streptococci (*Streptococcus agalactiae*), and *Listeria monocytogenes*. Bacterial meningitis in children older than 1 month and under the age of 4 years is caused mainly by *Haemophilus influenzae*, *Neisseria meningitidis*, and *S. pneumoniae*, with the risk for developing meningitis by age 5 about 1 in 1500.[6] Meningitis in older children and adults is usually caused by *N. meningitidis* and *S. pneumoniae*;[7] however, different bacterial agents are often involved in patients with cancer.[8] In a survey[7] of bacterial meningitis from 1977 to 1981, *H. influenzae* was found to be the most frequent cause of bacterial meningitis (48.3%), with 19.6% of the reported cases due to *N. meningitidis* and 13.3% due to *S. pneumoniae*. Meningitis caused by *H. influenzae* was most frequently seen in children between the ages of 6 months to 3 years, with peak activity in the spring and fall,[9] whereas *N. meningitidis* and *S. pneumoniae* were shown to occur primarily during the winter. *L. monocytogenes* is often associated with neonates or immunocompromised patients.[10] Meningitis caused by Gram-negative bacteria other than *H. influenzae* is often acquired nosocomially or as a result of head trauma and neurosurgical procedures.[10]

II. LABORATORY DIAGNOSIS OF BACTERIAL MENINGITIS

The life-threatening nature of bacterial meningitis requires that rapid and accurate laboratory diagnosis be performed on cerebrospinal fluid (CSF) to determine the extent of the inflammatory response and to provide, if possible, a specific etiology. Culture of CSF for isolation and identification of the etiological agent is required in all cases of suspected meningitis. Pretreatment with antibiotics often makes isolation difficult. In such cases, successful isolation may be achieved by using filtration methods or resins to absorb antimicrobials in the CSF.[11]

Chemical indicators for the presence of purulent meningitis form part of the routine laboratory workup, with low glucose and elevated protein CSF values consistent with a diagnosis of bacterial meningitis. Microscopic examination of CSF reveals a predominance of polymorphonuclear leukocytes, with total white blood cell (WBC) counts ranging from ≥500 to 10,000 cells per millimeter.[3,12]

Although a variety of laboratory tests are available to aid in rapid emergency diagnostic studies, microscopic examination of CSF by Gram stain is still the primary diagnostic test for preliminary diagnosis of purulent meningitis. Properly done, the Gram stain will detect microorganisms in about 70% of cases.[1,12,13] Acridine orange fluorescent staining of CSF has been shown to enhance detection of organisms in CSF;[14] however, positive smears must be Gram stained to determine Gram-reactivity.[7] False-negative Gram stains are common in partially

TABLE 1
Laboratory Tests for Detection of Bacterial Meningitis in Cerebrospinal Fluid

Conventional
- Chemistries: protein and glucose
- Microscopic: cell counts
  - Gram stain
  - Acid fast stain
  - Acridine orange stain
  - India ink preparation

Nonspecific
- *Limulus* amoebocyte lysate test (Gram-negative only)
- Lactic acid assay
- Lactic dehydrogenase isoenzymes
- Muramidase (lysozyme) assay
- Elastase-α_1-proteinase inhibitor
- C-reactive protein

Specific: antigen detection
- Quellung (capsular swelling) reaction
- Counter immunoelectrophoresis
- Staphylococcal coagglutination
- Latex particle agglutination
- ELISA

treated cases of meningitis or in cases in which the bacterial count is less than 1000 organisms per millimeter CSF.[15]

In addition to the conventional Gram stain of CSF, a number of other tests have been developed to aid in rapid diagnosis of bacterial meningitis (Table 1). As shown, some of these tests can provide specific information relative to the organism causing the meningitis, i.e., antigen detection methods, while others simply offer nonspecific, but diagnostic, information indicating the presence of bacterial pathogens in the CSF.

Nonspecific methods for rapid detection of bacterial meningitis can yield important diagnostic data. CSF lactic acid levels have been shown to be elevated in bacterial[16,17] and fungal[18] meningitides. CSF lactate levels follow the course of the clinical disease, with intermediary levels in partially treated cases and lower levels (normal) in fully treated cases or in cases with no bacterial meningitis. False-positive tests can occur in patients with central nervous system ischemia, subarachnoid hemorrhage, neoplasms, and closed-head injuries.[17] Lactic acid dehydrogenase (LDH) concentrations are also markedly increased during episodes of bacterial meningitis.[15,19] Additionally, specific LDH isoenzymes (LDH 5) are elevated in bacterial meningitis, whereas in viral meningitis the LDH isoenzyme pattern is close to that of normal or may show increases in levels of LDH 4. As with lactic acid levels, increases in LDH CSF concentrations may occur due to a number of conditions other than meningitis. C-reactive protein assay has been used to differentiate bacterial from nonbacterial meningitis.[20] This protein increases in concentration during central nervous system inflammatory disease; however, its usefulness in an acute setting may be of limited value.[21] Muramidase CSF concentrations have been shown to rise significantly in bacterial meningitis, returning slowly to normal levels during resolution of the disease.[22] Accurate diagnosis with muramidase was possible when comparing patients with meningitis to those in the aseptic and normal groups, with mean muramidase concentrations in bacterial meningitis far above those determined for the latter two groups. The authors[22] concluded that CSF muramidase activity could be used for providing a differential diagnosis of aseptic and bacterial meningitis and for following the course of the disease. Another nonspecific indicator for bacterial meningitis is elastase-α_1-proteinase, which binds and inactivates elastase extruded in the azurophilic granules of neutrophils. The protease inhibitor is markedly elevated during episodes of sepsis and bacterial meningitis[23] and is detected by enzyme-linked immunosorbent assay (ELISA). It was concluded that the proteinase inhibitor is useful for differentiating bacterial from nonbacterial meningitis and to determine adequacy of treatment.

Antigen detection methods have the advantage of providing specific information regarding the pathogen involved in the infectious process; however, limitations of these methodologies lie in the fact that only a limited number of organisms can be identified. Tests currently available for detection and specific identification of pathogens are shown in Table 1. The capsular swelling or Quellung test has long been used in the clinical microbiology laboratory. The test is easily performed and is done directly on spinal fluids containing sufficient numbers of organisms of appropriate cellular morphology. The test is largely limited to detection of *S. pneumoniae and H. influenzae.*

Countercurrent immunoelectrophoresis (CIE) has been widely used for rapid identification of *H. influenzae*, *S. pneumoniae*, *N. meningitidis*, and group B streptococci. The limiting factor is the sensitivity and specificity of the antibody directed against the capsular polysaccharides of these species.[24] The assay requires special equipment and may take from 2 to 4 h before results are available. Staphylococcal coagglutination and latex particle agglutination, in contrast, provide a rapid and simple means to perform detection methods for bacterial antigens in CSF and other body fluids. These tests, like CIE, may be positive even when the Gram stain and culture are negative. Disadvantages lie in the limited numbers of CSF bacterial pathogens which can be detected and the cost of these assays compared to microscopic methods.[21,25,26] False-positive and false-negative reactions have been reported to occur.[24,27] Enzyme immunoassays (EIA) are ELISAs detecting capsular antigens. Although sensitive and requiring only dilute antisera, the assay requires 4 h to complete and is considerably more complicated to run than both coagglutination and latex agglutination tests.[21,24]

The foregoing discussion was presented to briefly outline the varied tests available for rapid diagnosis of bacterial meningitis. The focus of the present discussion is the *Limulus* amoebocyte lysate (LAL) test for the presence of the endotoxic moiety of the Gram-negative cell wall, a component unique to Gram-negative bacteria.[28] Performed on initial spinal fluids, the test exhibits both a high sensitivity and specificity for diagnosis of Gram-negative bacterial meningitis. The LAL assay offers an exciting approach for a rapid differential diagnosis of Gram-negative meningitis, with results available within 10 to 15 min (chromogenic assay). In the following sections, use of this assay for the detection of Gram-negative bacterial meningitis is reviewed.

III. LAL TEST FOR GRAM-NEGATIVE MENINGITIS

A. REVIEW OF LITERATURE

The potential for clinical diagnostic use of the LAL test was first described by Levin and co-workers[29-31] for the detection of Gram-negative bacteremia and/or endotoxemia. These studies demonstrated the LAL test of blood to be a sensitive, rapid, and reliable method for detection of endotoxin in the bloodstream. Endotoxin detection in blood was complicated by the presence of an inhibitor[30] blocking the reaction of endotoxin with LAL, and requiring chloroform extraction of endotoxin from plasma prior to LAL test. However, the potential for LAL as an adjunct test for rapid diagnosis of Gram-negative infection in body fluids was realized. Subsequently, Jorgensen et al.,[32] demonstrated the usefulness of the LAL test as a screen for rapid diagnosis of Gram-negative bacteriuria, again demonstrating the potential clinical diagnostic usefulness of this unique test.

Bacterial meningitis is a life-threatening infection with a high morbidity and mortality rate if not properly and promptly treated with appropriate antimicrobial agents. Rapid detection and

TABLE 2
LAL Test of Initial Cerebrospinal Fluids[a]

Diagnosis	No. of CSFs	LAL/Total (%)
Gram-negative meningitis	38	38/38(100)
Gram-positive meningitis[b]	6	0/6 (0)
Aseptic meningitis	18	0/18(0)
Meningeal leukemia	4	0/4 (0)
No meningitis	46	0/46(0)

[a] Data abstracted from Nachum et al., *N. Engl. J. Med.*, 289, 931, 1973.
[b] Includes 1 case of tubercular meningitis.

determination of the Gram-reactivity of the etiologic agent would allow for immediate institution of specific antibiotic therapy. To this end, Nachum and co-workers[33] evaluated CSF using the LAL test as a potential method for detection of Gram-negative bacterial meningitis in a pediatric population. A total of 210 CSF specimens were collected from 112 patients with 44 (39%) initial CSFs culture positive, with 38 of Gram-negative etiology (Table 2). All initial CSF specimens containing Gram-negative bacteria were observed to be LAL positive, with the majority (35/38) read as positive within 15 min of incubation at 37°C. In contrast, CSFs obtained from patients with Gram-positive bacterial meningitis, tubercular meningitis, aseptic meningitis, meningeal leukemia, and normal (no infective agent) CSF were uniformly LAL negative (Table 2). These data were of particular interest as they demonstrated the utility of this test for detection of a potentially fatal infection. Additionally, problems associated with LAL tests of blood[29-31] were not evident when the test was used to test CSF, i.e., no evidence of an inhibitor was found. The striking sensitivity of the LAL test in CSF was thought, in part, to be due to the small volume system (CSF) as compared to a significantly larger blood volume.[34]

In the same study,[33] results of Gram stain and LAL test on initial CSFs were compared. Of 38 specimens with culture-proven Gram-negative meningitis, 25 (66%) were visualized on Gram stain vs. a detection rate of 100% for the LAL test. In addition, post-treatment CSFs were assayed with LAL and results compared to Gram stain and culture. Of these specimens, 68% were shown to be LAL positive as compared to a culture-positive rate of 24% and 16% for organisms visualized on Gram stain. In summary,[33] it was concluded that the LAL test was a rapid, sensitive, and reliable screening test for diagnosis of Gram-negative bacterial meningitis.

The successful outcome of the study by Nachum et al.[33] was confirmed in a larger series of patients.[35,36] Ross et al.[35] evaluated the LAL test in 335 infants and children, with 37/38 cases of Gram-negative meningitis detected by the LAL test within 60 min incubation at 37°C. All cases of Gram-positive, aseptic, and no meningitis were LAL negative. Comparison of LAL test to Gram stain on initial CSF specimens again demonstrated the superiority of LAL test, with only 71% (27/38) positive Gram stains as compared to 97% for the LAL test. Of 34 cases of *H. influenzae* meningitis, 10 were reported to have been pretreated with antibiotics prior to admission; however, all CSFs obtained from these patients were observed to be LAL positive. A bedside LAL test kit was devised[35] and designed for use by house officers and medical students. Each kit consisted of positive and negative controls, plus a tube for actual patient CSF testing. Assay was performed both at bedside and in parallel in the laboratory. The procedure was reported to be practical and simple to implement with a 98% correlation between results obtained by laboratory personnel and individuals performing the test at bedside.[35]

Additional studies by various investigators further confirmed the reproducibility and reliability of the LAL test on CSF for rapid detection of Gram-negative bacterial meningitis[36-39] in both children and adults. However, use of the LAL test for detection of neonatal Gram-negative meningitis produced conflicting results.[40] In this large multi-institutional study, 307 CSF specimens, obtained from 84 neonates, were tested for the presence of endotoxin by the LAL test for neonatal Gram-negative meningitis. Positive LAL tests were observed in only 71% of initial CSF specimens with culture-proven Gram-negative meningitis and in 14% of CSF specimens with sterile cultures. These data led McCracken and Sarff[40] to conclude that the LAL test was not a sensitive screening procedure for detection of Gram-negative meningitis in a neonatal patient population. Reasons for failure of the LAL test to detect Gram-negative meningitis in this patient population are not clear, but may in part be related to interpretation of weak LAL reactions or the presence of grossly bloody CSF specimens, the latter of which might have a negative influence on the outcome of the test. Additionally, lack of sensitivity of the test may have been a result of the actual numbers of organisms in the CSF during active infection. In contrast to community-acquired meningitis with *H. influenzae* type b and *N. meningitidis*, nosocomially acquired Gram-negative bacilli, e.g., *E. coli*, have been shown to be present in low numbers in CSF during active disease[41,42] and, as such, may have contributed to the false-negative data reported by McCracken and Sarff.[40] In contrast, Dyson and Cassady[43] reported positive LAL tests in all initial CSF specimens from neonates suffering from Gram-negative bacterial meningitis (5 *E. coli* and 1 *Serratia marcescens*); whereas the LAL test of CSF obtained from 14 infants with Gram-negative bacteremia and/or bacteriuria were uniformly LAL nonreactive. Negative LAL tests on CSF from infants with Gram-negative bacteremia support experimental data showing that circulating endotoxin does not cross the blood-brain barrier.[44]

The possible value of the LAL test as a prognostic tool to aid in the management of bacterial meningitis was first investigated by Berman et al.,[36] who quantitated endotoxin in CSF from 232 patients with suspected meningitis. The LAL test was observed to be positive in all 86 patients with culture-proven Gram-negative meningitis and uniformly negative in 146 patients with Gram-positive, fungal, tubercular, and no infectious meningitis. Quantitation of CSF endotoxin was done by serial 2-fold dilutions,[45] with endotoxin concentrations on initial CSF specimens ranging from 4 ng to >2000 ng/ml. Levels of endotoxin were shown to be higher for patients with *H. influenzae* type b CSF infection than for patients with *N. meningitidis*. Initial CSF endotoxin concentrations dropped rapidly following institution of appropriate antimicrobial therapy. Within 24 h, endotoxin concentrations ranged from 125 ng/ml to undetectable (<0.25 ng/ml). These data were consistent with data reported by Feldman[41] showing a rapid drop of at least 100-fold in numbers of viable organisms per ml CSF following antibiotic therapy, as well as a significant drop in detectable bacterial antigen.[42] Initial endotoxin concentrations were not correlated to granulocyte counts, protein or glucose levels in CSF, or to the presence of a positive gram stain, nor was any relationship noted between endotoxin concentrations and severity of the clinical disease or development of postmeningitic sequelae and death.[36] However, it was concluded that persistence of endotoxin in CSF might suggest incomplete removal of Gram-negative bacteria from the CSF and thus provide a sensitive indicator for the efficacy of antimicrobial treatment.[36] In an earlier study,[33] the reappearance of endotoxin in CSF following treatment was correlated with recurrent meningeal infection. Ross et al.[35] reported similar data, showing that positive LAL tests became negative within 48 to 72 h after initiation of antimicrobial therapy, equating a rapid return to LAL negativity with a good prognosis; however, in those patients for whom the LAL test remained positive, the disease course was difficult, with increased morbidity.

In a later study, Jorgensen and Lee[37] reported that in meningitis caused by *H. influenzae* type b, endotoxin persistence was not correlated to the presence of viable organisms; on the other hand, when meningitis was caused by other Gram-negative bacteria and LAL tests were persistently positive, organisms were isolated from these CSFs and therapy was deemed inadequate. McCracken and Sarff[40] were able to show a direct relationship between presence, persistence, and concentrations of endotoxin in CSF and disease outcome in a neonatal population. They observed a significantly higher percentage of initial positive CSF LAL tests

(100%) in infants who died as compared to those infants with abnormal clinical outcomes (60%) and those whose disease course was normal with a favorable outcome (43%). In addition, severity and outcome of the disease were shown to be correlated to the kinetics of endotoxin disappearance from the CSF. A direct correlation was noted between outcome and duration of endotoxin in CSF, with a mean duration of 3.8 d for infants who died, as compared to a mean of 2.5 d for infants with an abnormal outcome and 0.7 d for infants who recovered normally. Also, both median and mean CSF endotoxin concentrations were observed to be lowest in normal survivors and highest in those infants with abnormal outcomes and in those infants who died. Based on these data, it was concluded[40] that the LAL test may serve as a useful adjunct to other routine tests performed on CSF when bacterial meningitis is suspected.

To further assess the prognostic value of LAL test in CSF, Dwelle et al.[46] screened a total of 1503 CSF specimens using a *Limulus* lysate microassay (slide test). Results were compared to a number of clinical and laboratory variables. CSF endotoxin was quantitated and ranged from 0.1 ng/ml to 512 μg/ml. The clinical relevance of endotoxin CSF concentrations was shown by a significant correlation of seizures to initial endotoxin concentrations of ≥150 ng/ml. Increased morbidity was correlated to CSF endotoxin concentrations of ≥400,000 ng/ml, with 3 deaths noted in this group of patients. A correlation of CSF endotoxin concentration to peripheral leukocytosis was also observed, with concentrations of <3000 ng/ml associated with leukocytosis and endotoxin levels of ≥3000 ng/ml associated with neutropenia; however, other parameters usually associated with endotoxemia were not significantly correlated. These investigators found the LAL test to have prognostic value as a sensitive, specific screen for Gram-negative bacterial meningitis.

Methods used for detection of Gram-negative bacterial meningitis have varied among investigators. Initial studies[33,35-40,43,44,47,48] were done in the traditional manner for scoring a positive test,[34] i.e., incubation of equal volumes of CSF and LAL at 37°C for 1 to 4 h, with the test read by inverting the tube 180° and observing for the presence of a solid gel or by noting a marked increase in viscosity and turbidity. Less traditional gelation methods have been used, e.g., a microassay first introduced by Frausch[49] and modified for LAL of CSF.[25,46] In this test, CSF and LAL are mixed on pyrogen-free glass slides and incubated at 37°C for 60 min in a moist chamber. Positive tests are scored by tilting the slide and observing for gelation. A spectrophotometric method[50] has been described for LAL test of CSF using the Abbott Quantum bichromatic spectrophotometer together with specifically formulated LAL for turbidometric analysis. All cases of Gram-negative bacterial meningitis were successfully detected by this method; in contrast, patients with Gram-positive, fungal, or no infective meningitis were routinely shown to be LAL negative.

Chromogenic substrates[51] for the LAL test have been used by a number of investigators for rapid and reliable assay of endotoxin in CSF.[52-55] Advantages of the chromogenic test are that rapid color development (10 to 13 min), due to the relatively large concentrations of endotoxin that may be present in initial samples of CSF,[36,40,46] can be read visually (yellow color) or provide quantitative data by reading spectrophotometrically at O.D. 405 nm. Furthermore, the chromogenic test obviates clot formation as a positive end point, avoiding problems associated with clot dissolution by inadvertent rough handling of the tube during incubation or reading.[56] Thomas et al.,[54] described enhancement of chromogenic activity by heating CSF in a boiling water bath for 10 min, thus increasing endotoxin concentrations over those found for unheated CSF. The marked increase (up to fivefold) was thought to be a result of destruction of cellular proteases and activated coagulation factors interfering in the test.

Inagabi et al.,[57] have introduced a new chromogenic reagent (Endospecy) that is free of (1–3)-β-D-glucan (factor G), and claimed to be specific for endotoxin. Endospecy,[57] used to quantitate endotoxin in CSF obtained from neonates and infants, detected low levels of endotoxin (1.1 pg/ml) in both patient groups with no evidence of infective meningitis. Patients with Gram-negative bacterial meningitis were shown to have greatly elevated levels of endotoxin by this method. The detection of low levels of endotoxin in this study group (normal concentration of CSF endotoxin was shown to be ≤3 pg/ml) with no meningitis is of interest and has not been previously reported. It would appear that the sensitivity of this test is somewhat greater than that reported for other chromogenic LAL tests in which the chromogenic substrate S-2423 is used (Kabi Vitrum, Sweden) and may account for what might appear to be false-positive tests in otherwise normal CSFs. Further work with Endospecy needs to be performed to confirm these data. Background endotoxin levels in blood have been reported[58,59] and can be traced to leakage from the gastrointestinal tract;[60] however, crossing of endotoxin from the bloodstream into the spinal space does not appear to a likely mechanism[44,60] for background endotoxin CSF levels.

Saubolle[52] and Saubolle and Jorgensen[53] used the chromogenic LAL test for rapid detection of Gram-negative bacterial meningitis. In both studies chromogenic LAL tests were positive in all cases of community-acquired meningitis with *H. influenzae* type b and *N. meningitidis*; however, in nosocomially acquired meningitis caused by other Gram-negative bacilli, chromogenic LAL test sensitivity was only 67%. This apparent lack of sensitivity was speculated to be due to low numbers of bacteria present in the CSF,[52,53] and in the cases reported, all chromogenic LAL negative, culture-positive CSFs were shown to have colony counts of ≤50 colony forming units (CFU)/ml and often ≤10 CFU/ml.[53] Overall, sensitivity and specificity of LAL tests were 97% and 99%, respectively.[53]

Irrespective of the method used for detection, the LAL test of CSF has been shown to be both sensitive and specific for rapid detection of Gram-negative bacterial meningitis when compared to culture. The major drawbacks of the LAL test are the inability to detect infection of the CSF with microorganisms other than Gram-negative bacteria and the inability to identify the infecting Gram-negative organism. Nevertheless, the LAL test of CSF has potential, not only for rapid diagnosis of Gram-negative bacterial meningitis, but also as a prognostic indicator for efficacy of antimicrobial therapy and possible outcome of the disease.[61]

B. EFFICACY OF LAL SCREENING FOR GRAM-NEGATIVE BACTERIAL MENINGITIS

Efficacy and usefulness of the LAL test for detection of Gram-negative bacterial meningitis was initially demonstrated by Nachum et al.[33] Subsequent studies have confirmed the usefulness of the test as a rapid, sensitive, and reliable diagnostic tool for detection of Gram-negative bacterial meningitis, as summarized in Table 3. Out of a total of 692 culture-positive CSFs, only 52 (8%) were not detected by the LAL test. In contrast, false-positive LAL tests for confirmed cases of Gram-positive bacterial meningitis were only 7/210 (3%), for a false-positive rate of 3%. Further demonstration of the usefulness of the LAL test is shown in Table 4, in which an overall sensitivity and specificity has been calculated for a total of 4884 reported CSFs assayed. Of these, only 17 false-positive tests have been documented for a calculated specificity of >99%. Closer analysis of the false-positive tests shows that 3/17 were positive only after chloroform extraction of CSF obtained from 1 patient with pneumococcal meningitis and 2 from patients with streptococcal septic arthritis.[47] Reasons for these false-positive LAL tests were speculated to be related to the high white cell count (>10,000/mm^3) in the CSF, a correlation previously reported in plasma samples with elevated neutrophil counts[62] and joint fluids with elevated WBC counts.[63] However, a relationship between high WBC and positive LAL tests has not been confirmed or reported by other workers. Dyson and Cassady[43] described three false-positive LAL tests in their study. Each of these false-positive tests was scored as a 1+ reaction, a reaction which they described as a slight flocculation with no change in viscosity or turbidity following a 30-min incubation at 37°C. It is doubtful that the appearance of slight flocculation would be routinely reported as positive. Thomas et al.,[54] reported a false-positive test from culture-negative CSF obtained from a patient with clinically diagnosed Waterhouse-Friedricksen syndrome. Blood cultures were similarly negative for *N. meningitidis*; however, positive LAL tests in the absence of organisms isolated from CSF and blood would suggest that this patient

TABLE 3
Summary of LAL Test of Initial CSFs for Detection of Gram-Negative Meningitis

No. of CSFs tested	No. of LAL-positive CSFs/no. tested (%) GN meningitis	GP meningitis[a]	Ref.
112	38/38 (100)	0/6 (0)	33
335	37/38 (97)	0/13 (0)	35
232	86/86 (100)	0/23 (0)	36
305	61/61 (100)	0/14 (0)	37
307	60/87 (71)	—	40
208	5/5 (100)	0/4 (0)	43
3	3/3 (100)	—	38
25	24/25 (96)	0/7 (0)	39
81	8/8 (100)	1/6 (17)	47
53	4/4 (100)	0/1 (0)	54
103	11/11 (100)	0/4 (0)	50
146	49/53 (92)	0/13 (0)	52
1503	37/38 (97)	—	46
138	7/8 (88)	0/5 (0)	48
1054	136/142 (96)	0/59 (0)	53
53	3/3 (100)	—	56
155	37/48 (77)	6/44 (14)	25
62	25/25 (100)	0/11 (0)	55
9	9/9 (100)	—	68
Total			
4884	640/692(92)	7/210(3)	

[a] GN = Gram-negative; GP = Gram-positive.

TABLE 4
LAL Test of Initial CSFs

No. of CSFs tested	LAL Test[a] False-negative	False-positive	Sensitivity (%)	Specificity (%)
4884	52	17	93.0	99.4

[a] Compared to cultures for Gram-negative bacteria.

may have received prior antimicrobial treatment. The largest number of false-positive CSF LAL tests were reported by Hoban et al.,[25] who observed six such tests from CSFs growing out Gram-positive organisms and two culture-negative specimens. Reasons for the large number of false-positive LAL tests are not clear.

False-negative LAL tests (52 out of a total of 692 cases of Gram-negative bacterial meningitis) have been reported (Table 3) for an overall ability of the LAL test to detect 92% of these cases. The majority of these false-negative LAL tests have been reported by two groups of investigators.[25,40] McCracken and Sarff's[40] CSF studies involved a neonatal patient population. In this population meningitis is a nosocomially acquired disease caused by organisms other than *H. influenzae* type b and *N. meningitidis*. Nosocomially acquired CSF bacterial pathogens are present in counts of a magnitude possibly not detectable by the LAL test,[41,52,53] as opposed

TABLE 5
LAL Test of CSF by Bacterial Pathogen Isolated[a]

Organism	No. tested	No. LAL positive (%)
Gram-negative bacteria		
H. influenzae type b	347	345 (99)
N. meningitidis	73	73 (100)
E. coli	22	18 (82)
Klebsiella-Enterobacter-Serratia sp.	4	4 (100)
Other *Enterobacteriaceae*	10	8 (80)
Pseudomonas sp.	9	8 (89)
Other Gram-negative bacilli	8	8 (100)
Total	473	464 (98)
Gram-positive bacteria		
S. pneumoniae	87	2 (2)
Group B streptococcus	26	1 (4)
Other streptococci	3	1 (33)
Staphylococcus aureus	6	0 (0)
L. monocytogenes	18	1 (6)
Mycobacterium tuberculosis	13	0 (0)
Total	153	5 (3)
Other		
Fungi	14	0 (0)
Aseptic and viral	56	2 (1)
Total	50	2 (4)

[a] Data abstracted from References 25, 33, 35 to 40, 43, 47, 48, 50, 52 to 55, and 61.

to bacterial counts in CSF for community-acquired meningitis in which large numbers of organisms have been shown to be present in initial CSF specimens.[41] Hoban et al.,[25] reported a LAL detection rate of only 77%. Of the 48 culture-positive CSFs, 37 grew out *H. influenzae* type b and *N. meningitidis*; the other 11 cultures were caused by *E. coli*, *Serratia marcescens*, *Klebsiella pneumoniae*, and *Acinetobacter lwoffi*. Unfortunately, no breakdown by organism was given for the 11 false-negative LAL tests. Thus, it is not possible to determine if the false-negative LAL tests may have been due to nosocomially acquired microorganisms. Also, the LAL sensitivity was not stated, making assessment of these data difficult.

In Table 5, efficacy of LAL test is given by organism or groups of organisms. The combined sensitivity of the LAL test for detection of *H. influenzae* type b and *N. meningitidis* is 99.5%, data which are in agreement with conclusions reached by Saubolle[52] and Saubolle and Jorgensen[53] regarding community-acquired meningitis. In contrast, detection by the LAL test of nosocomially acquired meningitis caused by Gram-negative bacilli was shown to be only 86.8%. Overall specificity of the test was shown to be 99.4% (Table 4). Of 157 Gram-positive and 4 fungal meningitides, 7 false-positives (4%) were reported; similarly, of 156 reported viral and aseptic meningitides, 2 false-positive LAL reactions were observed.

Rapid diagnosis of bacterial meningitis by Gram stain has been a generally accepted method for demonstration of microorganisms before culture results are available.[7] Several investigators have compared Gram stain data obtained from initial CSF specimens to LAL detection of Gram-negative bacterial meningitis. As shown in Table 6, detection of Gram-negative bacterial

TABLE 6
Comparison of LAL Test and Gram Stain on Initial CSFs

No. of GN[a] meningitis	No. of correct Gram stains (%)	No. of LAL positive (%)	Ref.
38	25(66)	38(100)	33
38	27(71)	37(97)	35
61	44(72)	61(100)	37
6	3(50)	6(100)	43
45	28(62)	45(100)	52
8	6(75)	7(88)	48
106	86(81)	106(100)	53
Total			
302	219(73)	300(99)	

[a] GN = Gram-negative.

meningitis by the LAL test is far in excess of Gram stain of CSF (99% vs. 73%, respectively). The overall detection rate by Gram stain of 73% is in accord with other published data on the efficacy of Gram stains for demonstration of organisms on initial CSF specimens.[1,13]

Antigen detection methods for rapid diagnosis of bacterial meningitis are available for common bacterial pathogens causing meningitis.[24] These methods detect bacterial polysaccharide antigens by coagglutination, latex particle agglutination, and CIE. Several investigators[25,48,53,55] have compared these antigen detection methods to culture, LAL test, and Gram stain for rapid diagnosis of Gram-negative bacterial meningitis. With the exception of the study by Hoban et al.,[25] detection by the LAL test was shown as the most sensitive method for detecting *H. influenzae* type b and *N. meningitidis* in CSF; however, antigen detection methods have a distinct advantage over the LAL test, as specific pathogens can be identified and Gram-positive pathogens can also be detected. However, the cost of using antigen detection methods is considerably greater compared to either Gram stain or the LAL test.[25]

IV. CONCLUSIONS AND FUTURE TRENDS

As this discussion indicates, the LAL test of CSF for detection of Gram-negative bacterial meningitis is one of the most successful clinical diagnostic applications of the test. The severe nature of bacterial meningitis requires that rapid laboratory diagnosis be available for both Gram-positive and Gram-negative bacterial pathogens commonly associated with meningitis. Nevertheless, the LAL test of CSF can play an important role in the laboratory diagnosis of Gram-negative bacterial meningitis as an adjunct test and may, in certain instances, provide a diagnosis not always available by CSF cell count or Gram stain. Elevated CSF WBC counts are consistent with a diagnosis of bacterial meningitis. However, culture-positive CSF specimens with minimal pleocytosis[64] or normocellular CSF[64-66] have been reported. Similarly, CSF Gram stain data are not always reliable.[13,67] As shown in Table 6, LAL test of CSF is superior in its sensitivity to the Gram stain in detecting Gram-negative organisms in CSF. A clinical corollary to this is patients who present partially treated with antibiotics, and as a result, may have greatly reduced CSF organisms, impairing visualization of organisms on Gram stain,[42,68] and making a definitive diagnosis difficult. As shown,[33,35,37] the LAL tests may be positive following 24 to 48 h of antibiotic treatment, whereas Gram stain of these CSFs rarely demonstrate organisms. In such cases, the LAL test can provide a retrospective diagnosis.

Antigen detection methodologies can provide rapid detection of common CSF bacterial pathogens. Routine use of these tests may not be cost-effective[21,25,26,53] as compared to Gram stain and the LAL test. The low cost of the LAL test,[25,53] together with its high sensitivity and specificity (Table 4), make it a practical test for routine laboratory use. A positive LAL test would dictate that only Gram-negative profiles be performed; a negative LAL test and high polymorphonuclear leukocyte counts would dictate that Gram-positive antigen tests be done. Used in this manner, the LAL test of CSF could greatly reduce the cost of performing antigen detection tests.[53]

The prognostic potential of the LAL test on CSF has been discussed. Once a Gram-negative infection is established, the LAL test could be used to monitor efficacy of antibiotic treatment. Rapid resolution of the infection has been shown to be closely followed by rapid clearance of endotoxin from CSF.[33,36,68] In contrast, direct correlations have been made between duration of positive LAL results and outcome of the episode.[36,40] In addition, Dwelle et al.[46] have correlated CSF endotoxin concentrations with clinical and laboratory variables and found the test to have prognostic potential and to be a cost-effective screening test for Gram-negative bacterial meningitis.

Siegel and Nachum[61] have summarized possible indications for clinical use of the LAL test on CSF, including (1) rapid detection of Gram-negative bacterial meningitis on initial CSF specimens; (2) monitor efficacy of antibiotic therapy; and (3) detection of recurrent bacterial CSF infection following treatment. Additionally, the high sensitivity and specificity of the test demands that further consideration be given to this clinical diagnostic application of LAL test. Ease of performance, especially with chromogenic substrates, speed, low cost, and prognostic features suggest that the LAL test of CSF could easily be used in the laboratory or emergency room as a highly effective screening test for Gram-negative bacterial meningitis.

REFERENCES

1. **Carpenter, R. R. and Petersdorf, R. G.,** The clinical spectrum of bacterial meningitis, *Am. J. Med.*, 33, 262, 1962.
2. **Klein, J. O., Feigin, R. D., and McCracken, G. H., Jr.,** Report on the task force on diagnosis and management of meningitis, *Pediatrics*, 78, 959, 1986.
3. **Kaplin, S. L. and Feigin, R. D.,** Clinical investigations, prognostic factors and diagnosis of bacterial meningitis, in *Bacterial Meningitis*, Sande, M., Smith, A. L., and Root, R. K., Eds., Churchill Livingstone, New York, 1985, 83.
4. **Rennie, C.,** Meningitis, in *Emergency Medicine: A Comprehensive Study Guide*, Tintinalli, J. E., Rothstein, R. J., and Krome, R. L., Eds., McGraw-Hill, New York, 1985, 714.
5. Centers for Disease Control, Bacterial Meningitis and Meningococcemia: United States-1978, in *Morbid. Mortal. Wkly. Rep.*, 28, 277, 1979.
6. **Gold, R.,** Bacterial meningitis, *Am. J. Med.*, 75, 98, 1983.
7. **Ray, G. C., Wasilauskas, R. L., and Zabransky, R.,** Laboratory diagnosis of central nervous system infections, in *Cumitech*, Vol. 14, McCarthy, L. R., Ed., American Society for Microbiology, Washington, D.C., 1982.
8. **Chernik, N. L., Armstrong, D., and Posner, G.,** Central nervous system infections in patients with cancer, *Medicine*, 52, 563, 1973.
9. **Schleck, W. F., III, Ward, J. I., Band, J. D., Hightower, A., Fraser, D. W., and Broome, C. V.,** Bacterial meningitis in the United States, 1978 through 1981. The national bacterial meningitis surveillance study, *JAMA*, 253, 1749, 1985.
10. **Benson, C. A. and Harris, A. A.,** Acute neurological infections, *Med. Clin. North Am.*, 70, 987, 1986.
11. **Hansen, S. L., Hetmanski, J., and Stewart, B. J.,** Resin-process methods for improved isolation of organisms from blood and other body fluids, *Am. J. Med.*, 75, 31, 1983.
12. **Hyslop, N. E., Jr. and Swartz, M. N.,** Bacterial meningitis, *Postgrad. Med.*, 58, 120, 1975.
13. **Swartz, M. N. and Dodge, P. L.,** Bacterial meningitis — a review of selected topics. I. General clinical features, special problems and unusual meningeal reactions mimicking bacterial meningitis, *N. Engl. J. Med.*, 272, 725, 1965.

14. **Lauer, B. A., Reller, L. B., and Mirett, S.,** Comparison of acridine orange and Gram stains for detection of microorganisms in cerebrospinal fluid and other clinical specimens, *J. Clin. Microbiol.*, 14, 201, 1981.
15. **Martin, W. J.,** Rapid and reliable techniques for the laboratory detection of bacterial meningitis, *Am. J. Med.*, 75, 119, 1983.
16. **Bland, R. D., Lister, R. C., and Ries, J. P.,** Cerebrospinal fluid lactic acid levels and pH in meningitis, *Am. J. Dis. Child.*, 128, 1151, 1974.
17. **Jordan, G., Statland, B., and Halsted, C.,** Cerebrospinal fluid lactate in disease of the central nervous system, *Arch. Intern. Med.*, 143, 85, 1983.
18. **Body, B. A., Oneson, R. H., and Harold, D. A.,** Use of cerebrospinal fluid lactic acid concentration in the diagnosis of fungal meningitis, *Ann. Clin. Lab. Sci.*, 17, 429, 1987.
19. **Feldman, W. E.,** Cerebrospinal fluid lactic acid dehydrogenase activity. Levels in untreated and partially antibiotic-treated meningitis, *Am. J. Dis. Child.*, 129, 77, 1975.
20. **Corrall, C. J., Pepple, J. M., Moxon, E. R., and Hughes, W. T.,** C-reactive protein in spinal fluid of children with meningitis, *J. Pediatr.*, 99, 365, 1981.
21. **Dougherty, J. M. and Jones, J.,** Cerebrospinal fluid cultures and analysis, *Ann. Emer. Med.*, 15, 319, 1986.
22. **Ansari, A., Lipsey, A., and Nachum, R.,** Cerebrospinal fluid muramidase levels in meningitis, *J. Pediatr.*, 94, 752, 1979.
23. **Speer, C. P., Rethwilm, M., and Gahr, M.,** Elastase- α_1-proteinase inhibitor: an early indicator of septicemia and bacterial meningitis in children, *J. Pediatr.*, 111, 667, 1987.
24. **Kaplan, S. L.,** Antigen detection in cerebrospinal fluid — pros and cons, *Am. J. Med.*, 75, 109, 1983.
25. **Hoban, D. J., Witwicki, E., and Hammond, G. W.,** Bacterial antigen detection in cerebrospinal fluid of patients with meningitis, *Diagn. Microbiol. Infect. Dis.*, 3, 373, 1985.
26. **Welsh, D. F. and Hensel, D.,** Evaluation of Bactogen and Phadebact for detection of *Haemophilus influenzae* type b in cerebrospinal fluid, *J. Clin. Microbiol.*, 16, 905, 1982.
27. **Tilton, R. C., Dias, F., and Ryan, R. W.,** Comparative evaluation of 3 commercial products and counterimmunoelectrophoresis for detection of antigens in cerebro-spinal fluid, *J. Clin. Microbiol.*, 20, 23, 1984.
28. **Maitra, S. K., Nachum, R., and Pearson, F. C.,** Establishment of beta hydroxy fatty acids as chemical marker molecules for bacterial endotoxin by gas chromatography mass spectrophotometry, *Appl. Environ. Microbiol.*, 52, 510, 1986.
29. **Levin, J., Poore, T. E., Zauber, N. P., and Oser, R. S.,** Detection of endotoxin in the blood of patients with sepsis due to Gram-negative bacteria, *N. Engl. J. Med.*, 283, 1313, 1970.
30. **Levin, J., Tomasulo, P. A., and Oser, R. S.,** Detection of endotoxin in human blood and demonstration of an inhibitor, *J. Lab. Clin. Med.*, 75, 903, 1970.
31. **Levin, J., Poore, T. E., Young, N. S., Margolis, S., Zauber, N. P., Townes, A. S., and Bell, W. R.,** Gram-negative sepsis: detection of endotoxemia with the Limulus test. With studies of associated changes in blood coagulation, serum lipids, and complement, *Ann. Intern. Med.*, 76, 1, 1972.
32. **Jorgensen, J. H., Carvajal, H. F., Chipps, B. E., and Smith, R. F.,** Rapid detection of Gram-negative bacteriuria by use of the *Limulus* endotoxin assay, *Appl. Microbiol.*, 26, 38, 1973.
33. **Nachum, R., Lipsey, A. L., and Siegel, S. E.,** Rapid detection of Gram-negative bacterial meningitis by the *Limulus* lysate test, *N. Engl. J. Med.*, 289, 931, 1973.
34. **Siegel, S. E. and Nachum, R.,** Use of the Limulus lysate assay (LAL) for detection and quantitation of endotoxin, in *Perspectives in Toxicology*, Bernheimer, A. W., Ed., John Wiley & Sons, New York, 1977, 61.
35. **Ross, S., Rodriquez, W., Controni, G., Korengold, G., Watson, W., and Khan, W.,** Limulus lysate test for Gram-negative bacterial meningitis, *JAMA*, 233, 1366, 1975.
36. **Berman, N. S., Siegel, S. E., Nachum, R., Lipsey, A., and Leedom, J.,** Cerebrospinal fluid endotoxin concentrations in Gram-negative bacterial meningitis, *J. Pediatr.*, 88, 553, 1976.
37. **Jorgensen, J. H. and Lee, J. C.,** Rapid diagnosis of Gram-negative bacterial meningitis by the *Limulus* endotoxin assay, *J. Clin. Microbiol.*, 7, 12, 1978.
38. **Butler, T., Levin, J., Cu, D. Q., and Walker, R. I.,** Bubonic plague: detection of endotoxemia with the Limulus test, *Ann. Intern. Med.*, 79, 642, 1973.
39. **Clumeck, N., Lauwers, S., and Butzler, J. P.,** Limulus test and meningitis, *Br. Med. J.*, 1, 777, 1977.
40. **McCracken, G. H., Jr. and Sarff, L. D.,** Endotoxin in cerebrospinal fluid. Detection in neonates with bacterial meningitis, *JAMA*, 235, 617, 1976.
41. **Feldman, W. E.,** Concentrations of bacteria in cerebro-spinal fluid of patients with bacterial meningitis, *J. Pediatr.*, 88, 549, 1976.
42. **Feldman, W. E.,** Relationship of concentrations of bacteria and bacterial antigen in cerebrospinal fluid to prognosis in patients with bacterial meningitis, *N. Engl. J. Med.*, 296, 433, 1977.
43. **Dyson, D. and Cassady, G.,** Use of *Limulus* lysate for detecting Gram-negative neonatal meningitis, *Pediatrics*, 58, 105, 1976.

44. **Trippodo, N. C., Jorgensen, J. H., Priano, L. L., and Traber, D. L.,** Cerebrospinal fluid levels of endotoxin during endotoxemia, *Proc. Soc. Exp. Biol. Med.*, 143, 932, 1973.
45. **Reinhold, R. B. and Fine, J.,** A technique for quantitative measurement of endotoxin in plasma, *Proc. Soc. Exp. Biol. Med.*, 137, 334, 1971.
46. **Dwelle, T. L., Dunkel, L. M., and Blair, L.,** Correlation of cerebrospinal fluid endotoxinlike activity with clinical and laboratory variables in Gram-negative bacterial meningitis in children, *J. Clin. Microbiol.*, 25, 856, 1987.
47. **Tuazon, C. U., Perez, A. A., Elin, R. J., and Sheagren, J. N.,** Detection of endotoxin in cerebrospinal fluid and joint fluids by Limulus assay, *Arch. Intern. Med.*, 137, 55, 1977.
48. **Khurana, C. M. and Deddish, P. A.,** Comparison of results of Limulus amebocyte lysate, counterimmunoelectrophoresis, and Gram stain on spinal fluids of patients with suspected meningitis, *Curr. Ther. Res.*, 41, 604, 1987.
49. **Frausch, P.,** Slide test as a micromethod of a modified *Limulus* endotoxin test, *J. Pharm. Sci.*, 63, 808, 1974.
50. **Jorgensen, J. H., Redding, J. S., Reichler, A. A., and Spencer, H. J.,** Spectrophotometric Limulus amoebocyte lysate assay for rapid detection of Gram-negative bacterial meningitis or significant bacteriuria, *Abstracts, Interscience Conference on Antimicrobials and Chemotherapy,* No. 118, American Society for Microbiology, Washington, D. C., 1984, 110.
51. **Iwanaga, S., Morita, T., Harada, T., Nakamura, S., Niwa, M., Takada, K., Kimura, T., and Sakakibara, S.,** Chromogenic substrates for horseshoe crab clotting enzyme. Its application for the assay of bacterial endotoxins, *Hemostasis*, 7, 183, 1978.
52. **Saubolle, M. A.,** Chromogenic *Limulus* amebocyte lysate assay as an aid in the diagnosis of meningitis, in *Progress in Clinical and Biological Research*, Vol. 189, Ten Cate, J. W., Bueller, H. R., Sturk, A., and Levin, J., Eds., Alan R. Liss, New York, 1985, 369.
53. **Saubolle, M. A. and Jorgensen, J. H.,** Use of the *Limulus* amebocyte lysate test as a cost-effective screen for Gram-negative agents of meningitis, *Diagn. Microbiol. Infect. Dis.*, 7, 177, 1987.
54. **Thomas, L. L. M., tenCate, H., Sturk, A., Joop, K., and Ten Cate, J. W.,** Quantitative chromogenic endotoxin determination in cerebrospinal fluid, *Clin. Chim. Acta*, 127, 137, 1983.
55. **Saubolle, M. A., Barone, T., Blattman, A., and Ramos, A.,** Chromogenic *Limulus* amoebocyte lysate test for endotoxin in cerebrospinal fluid, *Abstracts, Am. Soc. Microbiol.*, C-120, 256, 1984.
56. **Nachum, R. and Berzofsky, R. N.,** Chromogenic *Limulus* amoebocyte lysate assay for rapid detection of Gram-negative bacteriuria, *J. Clin. Microbiol.*, 21, 759, 1985.
57. **Inagabi, M., Nakajima, M., Ando, Y., Takashima, S., Takeshita, K., Tanaka, S., Tamura, H., and Obayashi, T.,** Measurement of endotoxin in cerebrospinal fluid from neonates and infants using a new endotoxin-specific chromogenic test, *Clin. Chim. Acta*, 175, 315, 1988.
58. **Berger, D., Marzinzig, E., Marzinzig, M., and Beger, H. G.,** Quantitative endotoxin determination in blood — chromogenic modification of the Limulus amebocyte lysate test, *Eur. Surg. Res.*, 20, 128, 1988.
59. **Pearson, F. C., Duabczak, J., Weary, M., Bruszer, G., and Donohue, G.,** Detection of endotoxin in the plasma of patients with Gram-negative bacterial sepsis by the *Limulus* amoebocyte lysate assay, *J. Clin. Microbiol.*, 21, 865, 1985.
60. **Dascombe, M. J. and Milton, A. S.,** Study on the possible entry of bacterial endotoxin and prostaglandin E2 into the central nervous system from the blood, *J. Pharmacol.*, 66, 565, 1979.
61. **Siegel, S. E. and Nachum R.,** Indications for the use of the Limulus lysate assay in bacterial meningitis, in *PProgress in Clinical Biological Research,* Vol. 29, Cohen, E., Ed., Alan R. Liss, New York, 1979, 245.
62. **Elin, R. J., Robinson, R. A., Levine, A. S., and Wolff, S. M.,** Lack of clinical usefulness of the Limulus test in the diagnosis of endotoxiemia, *N. Engl. J. Med.*, 293, 521, 1975.
63. **Elin, R. J., Knowles, R., Barth, W. F., and Wolff, S. M.,** Lack of specificity of the Limulus lysate test in the diagnosis of pyogenic arthritis, *J. Infect. Dis.*, 137, 507, 1978.
64. **Moore, C. M. and Ross, M.,** Acute bacterial meningitis with absent or minimal cerebrospinal fluid abnormalities, *Clin. Pediatr.*, 12, 117, 1973.
65. **Onorato, I. M., Wormser, G. P., and Nicholas, P.,** Normal cerebrospinal fluid in bacterial meningitis, *JAMA*, 244, 1469, 1980.
66. **Smales, O. R. C. and Rutter, N.,** Difficulties in diagnosing meningococcal meningitis in children, *Br. Med. J.*, 1, 588, 1979.
67. **Feldman, W. E.,** Effect of prior antibiotic therapy on concentrations of bacteria in cerebrospinal fluid, *Am. J. Dis. Child.*, 132, 672, 1978.
68. **Arditi, M., Ables, L., and Yogev, R.,** Kinetics of endotoxin in CSF of children with Gram-negative bacterial meningitis, *Abstracts Interscience Conference on Antimicrobials and Chemotherapy,* No.6, American Society for Microbiology, Washington, D. C., 1988, 109.

Chapter 7

DETECTION OF GRAM-NEGATIVE BACTERIURIA

Ronald Nachum

TABLE OF CONTENTS

I. Introduction and Background 82

II. Review of Literature: Detection of Gram-Negative Bacteriuria by the LAL Test 82
 A. Gelation Endpoint Methods 83
 B. Spectrophotometric and Turbidometric Endpoint Methods 85
 C. Chromogenic Substrates 85

III. Clinical Diagnostic Usefulness of the LAL-Urine Screening Assays 87
 A. False-Negative LAL-Urine Assays 88
 B. False-Positive LAL-Urine Assays 89
 C. LAL-Urine Test: Comparison to Other Bacteriuria Screening Methods 91

IV. Conclusions and Future Trends 92

References 93

I. INTRODUCTION AND BACKGROUND

Laboratory diagnosis of urinary tract infection (UTI) constitutes a large part of the work load in clinical microbiology laboratories. UTIs may range from lower urinary tract disease, i.e., cystitis, to more serious infections of the upper urinary tract, i.e., pyelonephritis, which may progress to bacteremia, shock, hypotension, and kidney damage. UTI acquisition is usually by an ascending route, with microorganisms ascending from the urethra into the bladder, and possibly to the kidneys through the ureters.[1] As the focus of this presentation is the detection of clinically significant Gram-negative bacteriuria by *Limulus* amoebocyte lysate (LAL) test, discussion on the epidemiology and etiology of UTI is in the form of a brief overview; however, several excellent reviews are available for more detailed information.[2-5]

UTI occurs in all age groups, with prevalence rates age and sex dependent. Prevalence rates are highest in school age and adult females,[2,3] with a ratio of female to male of 30 or greater,[5] with declining prevalence rates in both newborns[6] and the elderly.[7] Symptomatic UTI in the female is most frequently observed in the 15 to 24 year old age range; however, asymptomatic UTI has been shown to be low in preteen females and increasing in prevalence to the age of 60.[8]

The majority of UTIs are bacterial in nature, with most infections caused by facultatively anaerobic Gram-negative bacteria;[5,9] however, Gram-positive bacteria also cause significant numbers of UTIs. Additionally, occasional infections are caused by anaerobic bacteria, mycoplasmas, fungi, viruses, and mycobacteria.

Laboratory diagnosis of UTI by conventional methods involves culture of viable bacteria from a midstream-voided urine specimen.[10] Quantitation is done by streaking appropriate media with 0.001 ml (calibrated loop method) such that the generally accepted value for diagnosis of clinically significant bacteriuria, i.e., $\geq 10^5$ colony forming units (CFU) per milliliter urine of a single pathogen,[11] can be easily made by counting colonies on the streak plate. This method has a turnaround time of at least 18 to 24 h, and up to 48 h may be required before bacterial counts are available. Less traditional methods have been introduced for rapid screening of bacteriuria, with the advantage of allowing the laboratory or physician office practice to eliminate culture-negative specimens. Both nonautomated and automated methods are available for rapid urine screening for UTI, the latter often requiring expensive equipment as detection of bacteriuria is done by various methodologies including photometry,[12-14] filtration,[15,16] and bioluminescence.[17,18] Nonautomated urine screening methods usually provide ease of operation, interpretation, and speed, with detection times of 2 to 30 min, with the exception of the dip-slide (Urocult) which requires overnight incubation.[19,20] Nonautomated methods include microscopy,[15,21-23] chemical methods,[24,25] and a disposable filtration method.[26] In addition, a number of published studies have demonstrated the usefulness of the LAL test for rapid diagnosis of clinically significant Gram-negative bacteriuria (Table 1).

II. REVIEW OF LITERATURE: DETECTION OF GRAM-NEGATIVE BACTERIURIA BY THE LAL TEST

In contrast to LAL testing of blood, cerebrospinal fluid (CSF), and endocervical and urethral exudates for detection of Gram-negative infectious disease,[27,28] clinical investigations utilizing the LAL test for detection of clinically significant Gram-negative bacteriuria have been relatively few in number. However, published data on the efficacy of LAL testing of urine have shown the assay to be as successful as documented for LAL-mediated detection of Gram-negative meningitis.[28-30] A compilation of currently available literature for rapid detection of Gram-negative bacteriuria by the LAL test is given in Table 1. For purposes of the following discussion, data in Table 1 has been divided according to the type of endpoint assay used for detection, i.e., gelation and/or increase in turbidity (spectrophotometric, turbidometric) and chromogenic.

TABLE 1
Summary: LAL-Urine Studies for Detection of Gram-Negative Bacteriuria

Test endpoint	No. urines tested	Assay time & temp.	No. positive LAL tests (%) ≥10^5 GNB/ml	No. positive LAL tests (%) ≥10^5 CFU/ml[a]	Ref.
Non-chromogenic gelation					
Jorgensen et al., 1973	209	60 min, 37°C	23/23 (100)	23/25 (92)	31
Jorgensen & Jones, 1975	1077	120 min, 37°C	159/161 (99)	175/203 (86)	32
Nachum & Shanbrom, 1981	190	15 min, 37°C	36/36 (100)	36/43 (84)	35
Krause-Hooyman & Komorowski, 1987	—	30 min, 25°C	93.8%	—	38
Spectrophotometric					
Jorgensen & Alexander, 1982	580	20 min, 35°C	78/83 (94)	78/92 (85)	47
Jorgensen et al., 1984	301	30 min, 37°C	32/34 (94)	—	48
Chromogenic					
Nachum & Berzofsky, 1985	324	5 min, 37°C	68/68 (100)	68/82 (82)	41
Nachum et al., 1986	1039	5 min, 25°C	74/86 (86)	74/101 (73)	52
Nurminen et al., 1988	311	13 min, 25°C	35/35 (100)	—	56
Luke et al., 1988					57
(1) 2-step format	93	12 min, 37°C	99.0%	—	
(2) 1-step format	16	20 min, 37°C	81.0%	—	

[a] Total Gram-positive, Gram-negative, and fungal.

A. GELATION ENDPOINT METHODS

LAL testing of urine was first described in 1973 by Jorgensen and co-workers.[31] They tested 209 urine specimens undiluted and diluted 1:10 to 1:1000 and reacted with equal volumes of lysate (sensitivity of 1 ng/ml to *Escherichia coli* 055:B5 endotoxin. Standard curves of *Limulus* activity and cell counts of at least 10^5 Gram-negative bacteria (GNB) per milliliter urine indicated that gelation, or a marked increase in viscosity, at urine dilutions of 1:100 or 1:1000 could be scored as a positive test; whereas positive LAL reactions obtained only from undiluted urine or urine diluted 1:10 were scored as negative, i.e., Gram-negative bacterial counts of $<10^5$. In this test, 23/23 patient urines with $>10^5$ GNB per milliliter were detected within the 60 min (37°C) incubation period established for the LAL test. From these data[31] it was concluded that *Limulus* testing of urine might be a useful screening test for patients with asymptomatic bacteriuria, providing the user with a simple, rapid, and objective test.

Subsequently, Jorgensen and Jones[32] compared results of the LAL test to direct Gram stain of urine for rapid and presumptive diagnosis of Gram-negative bacteriuria. Urine was diluted 1:100 in pyrogen-free saline and reacted with equal volumes of LAL (sensitivity = 5 ng/ml *E. coli* 055:B5 endotoxin) for 2 h at 37°C. Of 161 urines with counts of $>10^5$ GNB per milliliter, 159 (98.8%) were detected by the LAL test; in contrast, only 120/161 (74.5%) were detected by direct Gram stain. Comparison of data in which both Gram-negative and Gram-positive bacteriuria were tabulated were of interest in that the LAL test was still found to be more sensitive (86.2% vs. 69.5%, respectively), and specific (96.1% vs. 92.8%, respectively), than the direct Gram stain, even though the LAL test is nonreactive with Gram-positive bacteriuric urines.

Increasing concentrations of endotoxin have been shown to increase the rate at which LAL gelation takes place.[33,34] Using this direct relationship, Nachum and Shanbrom[35] evaluated a "time to gelation" method for rapid detection of $\geq 10^5$ GNB per milliliter in undiluted urine. Using a relatively insensitive LAL preparation (0.5 ng/ml *E. coli* EC-2 endotoxin), these investigators were able to show, in a series of preliminary experiments, that filter sterilized urine seeded with 1 to 2×10^5 GNB per milliliter caused LAL gelation within 10 to 15 min at 37°C. In contrast, Gram-negative counts of $<10^5$ resulted in gelation times of 20 min or greater. Urine seeded with $>10^5$ Gram-positive organisms per milliliter were LAL nonreactive. Based on these data, undiluted patient urine was reacted with equal volumes of LAL and incubated for 10 and 15 min (2 tubes) at 37°C, with 33/36 bacteriuric urines correctly detected. The three false-negative assays were subsequently shown to be a result of urine pH below the minimum for *Limulus* reactivity.[36] Adjustment of urine pH to 7.0 resulted in positive LAL tests and a 100% correlation with quantitative culture data. Of the 36 bacteriuric urines, 31 (86.1%) were LAL positive after only 10 min incubation at 37°C. The overall ability of the test to correctly differentiate urine specimens as containing $<10^5$ or $>10^5$ GNB per milliliter was 96.2%. It was observed that 4/6 urines with Gram-negative counts ranging from 5×10^4 to $<10^5$ were LAL positive within the 15 min time constraints of the test; however, all specimens with Gram-negative bacterial counts of 10^3 to $<5 \times 10^4$ were LAL negative, as were 7 urines with Gram-positive counts of $>10^5$. Two positive LAL tests were obtained from patient urine for which no growth ($<10^3$ CFU per milliliter) was observed. Both patients were subsequently shown to have gonococcal urethritis, suggesting that these two urine specimens may have contained either large numbers of gonococci and/or considerable amounts of cell-free gonococcal endotoxin, probably accounting for the positive LAL tests, especially in view of the fact that *Neisseria gonorrhoeae* endotoxin has been shown to have greater reactivity with LAL, as compared to endotoxins obtained from other common Gram-negative pathogens.[37] Based on their findings, these investigators found the "time to gelation" method to have several advantages for screening of urine, including a high degree of sensitivity and specificity, speed (10 to 15 min), ease of performance (no urine dilution required), and an easily interpreted endpoint.[35]

An interesting variation of the LAL gelation test for screening of Gram-negative bacteriuria has been described by Krause-Hooyman and Komorowski.[38] The test consists of a disposable device containing LAL which is rehydrated with buffer, followed by the addition of 3 drops of undiluted urine and incubation for 20 min at room temperature. At this time an indicator sphere is dropped into the test device, and the test is scored as positive if the indicator sphere floats on top of the reaction mixture; however, if the indicator sphere sinks to the bottom of the reaction mixture, a negative test is scored. Clinical tests demonstrated that this device correctly detected 93.8% of urines with Gram-negative bacterial counts of $>10^5$. Although the predictive value of a negative test was high (98.5%), the predictive value of a positive test was only 57.5%, considerably below that shown previously for other LAL screening tests for UTI.[39]

Miller et al.[40] have described a LAL gel endpoint method which might be used as an indicator for upper UTI, i.e., renal bacteriuria in polycystic kidney disease (PKD). The test was performed by reacting LAL (minimum sensitivity of 0.06 endotoxin units [Eu/ml]) with urine dilutions of 2,- 5,- and 50-fold. Out of 11 PKD urines assayed, positive LAL tests were obtained from urine dilutions of 1:2 (9/11), 1:5 (8/11), and 1:50 (3/11). To determine if the LAL reactive substance was endotoxin, LAL positive urines were heated to 100°C and retested. All such heat-treated urines remained LAL positive, indicating that LAL reactivity was endotoxin mediated. Since all urines were culture-negative, it was concluded that LAL testing of urine may be a sensitive test for renal Gram-negative bacteriuria in PKD patients.[40] However, Nachum and Berzofsky[41] have shown that culture-negative urine ($<10^3$ CFU/ml) obtained from female patients contained endotoxin concentrations ranging from 3.5 to 20 ng/ml. Of the 9 PKD patients[13] with positive LAL tests (7 males and 2 females), urines obtained from both female patients were LAL positive at dilutions of 1:5 and 1:50. It is possible that the urine endotoxin assayed may have entered the urine, either from resident urethral Gram-negative microorganisms (male or female) or from Gram-negative bacteria and/or free endotoxin from perineal or vaginal surfaces. Thus, endotoxin detected by the LAL test may not be representative of renal bacteriuria in PKD patients; however, use of the LAL test for detection of Gram-negative upper urinary tract disease is a novel use of the test and deserves further investigation and clinical trials.

B. SPECTROPHOTOMETRIC AND TURBIDOMETRIC ENDPOINT METHODS

Nonchromogenic spectrophotometric and/or turbidometric methodologies have been described for detection of endotoxin with the LAL test.[36,42-44] These methodologies offer greater resolution, objectivity, and increased sensitivity for LAL-endotoxin assay[45,46] as compared to the gel-clot endpoint assay. Utilization of LAL-spectrophotometric endpoint assay for detection of clinically significant Gram-negative bacteriuria have been described by Jorgensen and Alexander[47] and Jorgensen et al.[48] The advantage of spectrophotometric or kinetic-turbidometric assays is the ability to perform these tests on automated instrumentation,[45,46] producing reproducible and standardized data. Jorgensen and Alexander[45] used the Abbott MS-2 Microbiology System for performing large-scale automated LAL-endotoxin testing. Lysate preparations obtained from various commercial sources were readily adapted for use with the MS-2 system, with the added advantage of increasing the sensitivity of these preparations 60- to 250-fold, allowing for detection of as little as 1 pg endotoxin per milliliter.[45] In parallel with quantitative urine culture, MS-2 LAL-endotoxin automated assay was used to test 580 urines for Gram-negative counts of either $<10^5$ or $>10^5$/ml urine.[20] The assay was performed in special vials (Ampvettes) adapted for use in the MS-2 system, with 400 µl of a 1:2000 urine dilution reacted with 100 µl of LAL. The Ampvette was placed into the MS-2 analysis module and incubated 20 min at 35.5°C. Positive tests were scored when a change of 100 or greater MS-2 units occurred within any 1 to 5 min reading interval during the 20 min incubation period. The MS-2 automated system correctly identified 78/83 urines (94%) containing $>10^5$ GNB per milliliter 78 (94%). In addition, 10/25 (40%) of urines with Gram-negative counts of between 10^4 to 10^5 were detected by the MS-2 LAL system as were 3/426 urines with counts of $<10^4$ viable bacteria or no growth, constituting a false-positive rate of less than 1%. From these data, the investigators concluded that MS-2 LAL assay is less subjective and more sensitive than the gel endpoint method, having the distinct advantage of being able to perform large numbers of tests simultaneously.[45,47]

Turbidometric endpoints, as read on the Abbott UV Quantum bichromatic spectrophotometer, were used by Jorgensen et al.[48] as a rapid and objective method for differentiating urine specimens on the basis of $<10^5$ or $>10^5$ GNB per milliliter. Data similar to those reported for MS-2 LAL urine assay[20] were obtained with this automated system, with 32/43 (94%) of the cases successfully detected.

C. CHROMOGENIC SUBSTRATES

The introduction of chromogenic substrates[49] for detection and quantitation of endotoxin has simplified the LAL test and has increased the potential clinical diagnostic usefulness of the assay. Given the potentially large concentrations of endotoxin expected to be present in urine containing $>10^5$ Gram-negative urinary tract pathogens per ml and the large amounts of free endotoxin shown to be extruded by GNB into the surrounding menstrum,[36,50] the possibility of using a chromogenic LAL (CLAL) assay for rapid detection of Gram-negative bacteriuria was investigated by Nachum and Berzofsky.[41] Following a series of preliminary experiments, a 5-min urine-LAL incubation at 37°C was established to be sufficient for activation of the *Limulus* proclotting enzyme such that addition of the chromogenic substrate would result in rapid (30 s maximum) color development. Testing was performed at urine dilutions of 1:10 and 1:20, amounts necessary to quench the natural color of urine so as to avoid confusion in reading the chromogenic endpoint of the assay. Over 300 (324) urines were tested in parallel with quantitative urine culture.[10] Out of 68 specimens containing $>10^5$ GNB per milliliter 1 false-negative test was found (1:20 urine dilution); however, 9 false-positive LAL tests were identified, occurring at both urine dilutions and in the same specimens. Chloroform extracts[51] of the nine false-positive urines were all successfully retested, indicating that the original positive CLAL reactions were endotoxin-mediated and not the result of nonspecific activation of the proclotting enzyme. CLAL assay of urine was shown to be rapid (5 min) and easy to perform, with a high degree of sensitivity and specificity, resulting in high negative and positive predictive values.[41]

Nachum et al.[52] modified the CLAL testing of urine for use at room temperature. Room temperature assay for clinically significant Gram-negative bacteriuria was postulated to be feasible, as based on data[41] showing that large amounts (ranging from 200 to 2150 ng/ml) of endotoxin were present in Gram-negative bacteriuric urine. Using known culture-positive urines ($>10^5$ GNB per milliliter), optimal conditions for room temperature LAL testing were determined. Data obtained from these experiments showed that the assay was not sufficiently sensitive at room temperature, detecting only 60% of known culture-positive urines, in contrast to 100% when the assays were done at 37°. To obviate the lack of CLAL sensitivity at room temperature a surfactant, Tween 80 (polysorbate 80), shown to dissociate endotoxin into subunits,[53] (possibly exposing more active sites for LAL reactivity, i.e., lipid A)[54] was used as diluent at a concentration of 0.1%. Use of 0.1% Tween 80 for a urine dilution of 1:10 enhanced chromogenic *Limulus* activity, with 10/10 of the known culture-positive bacteriuric urines detected. A total of 1039 urine specimens were obtained from 664 obstetric patients for purposes of ascertaining the efficacy of room temperature CLAL assay and to document the rate of asymptomatic bacteriuria of pregnancy for the patient population tested. Of these, 74/86 (86%) were correctly detected by room temperature CLAL assay. Eleven false-positive tests were observed with 10/11 urines heavily contaminated, i.e., mixed Gram-positive flora suggesting that proper cleansing procedures had not been used prior to voiding.[10] The false-positive urine for which no organisms were isolated ($<10^3$) was obtained from a patient with endocervical gonorrhea and may have been responsible for the positive CLAL assay;[35] 12 false-negative tests were reported. Possible reasons for failure of room temperature CLAL assay to detect these bacteriuric urines is discussed later in this chapter. Predictive values were calculated for prevalence rates[55] of 5, 8, and 10%, with positive predictive values of 80.0, 87.5, and 89.9%, respectively, and negative predictive values of 99.4, 98.9, and 98.6%, respectively.

Spectrophotometric CLAL assay was evaluated by Nurminen et al.[56] for detection of Gram-negative bacteriuria. For assay, urine was diluted 1:10 in dissolving buffer containing 0.1% Tween 80 and reacted with LAL for 10 min at room temperature, followed by the addition of chromogenic substrate (Table 1). Assays were done in flat-bottomed microtiter plates. The reaction was stopped by the addition of acetic acid and tests read at an OD of 405 nm using an enzyme immunoassay reader. Positive tests were scored at OD_{405} readings between 0.5 to 1.0 or greater. All culture-positive urines were detected by spectrophotometric reading of microtiter plates. No false-positive assays were observed and Gram-negative counts of 10^4 to $<10^5$/ml were recorded as negative, i.e., OD_{405} of <0.5. Urine samples from male patients with gonococcal urethritis were all read as positive, data not dissimilar from that previously described for LAL assay of urine.[35,41,52] Advantages of spectrophotometric CLAL assay for detection of clinically significant Gram-negative bacteriuria[56] were ease, speed, and the ability to perform large numbers of tests simultaneously and objectively.

Rapid screening for Gram-negative urinary tract infection utilizing a *Limulus* dipstick procedure has been described by Luke et al.[57] For assay, two formats were presented, i.e., a one- and a two-step methodology. Parallel testing of the two-step format against 93 culture-positive urine specimens ($>5 \times 10^4$ GNB per milliliter) showed this form of the assay to correlate at a level

TABLE 2
Data Summary: Results of LAL-Urine Screening for Gram-Negative Bacteriuria

Test endpoint	Total urines tested	Total ≥105 GNB[a]/ml	LAL positive (%)	LAL false positive (%)	LAL false negative (%)
Gelation[b]	1450	221	219 (99.1)	26 (1.8)	2 (0.14)
Chromogenic	1674	189	177 (93.7)	20 (1.2)	12 (0.07)

[a] GNB: Gram-negative bacilli.
[b] Spectrophotometric endpoint data not included.

TABLE 3
Predictive Values for Detection of Significant Bacteriuria by the LAL Test

Patient urine[a]	% Sensitivity	% Specificity	% Predictive value	
			Positive	Negative
≥105 GNB[b]/ml				
Gelation	99.1	97.9	89.5	99.8
Chromogenic	94.0	98.7	90.4	99.2
≥105 Orgs/ml[c]				
Gelation	89.3	98.0	91.2	97.6
Chromogenic	84.2	98.6	92.8	97.3

[a] Total of 1450 urines tested by gel endpoint; 1674 by chromogenic assay.
[b] GNB: Gram-negative bacilli.
[c] Includes Gram-negative, Gram-positive, and fungal infections.

of 99% with the gel-clot test, with a negative predictive value of 100% as compared to culture results. The one-step format, although offering greater ease of testing, was shown to be considerably less sensitive than reported for the two-step format (81 vs. 94%), with a negative predictive value of 72 and 91%, respectively.

III. CLINICAL DIAGNOSTIC USEFULNESS OF THE LAL-URINE SCREENING ASSAYS

The discussion presented in Section II has shown *Limulus* assay of urine for detection of clinically significant Gram-negative bacteriuria to be one of the more successful applications of this unique assay, irrespective of endpoint used for differentiation of $<10^5$ or $>10^5$ GNB per milliliter urine. The overall success of LAL-urine assay is summarized in Tables 2 and 3, with data divided on the basis of nonchromogenic and chromogenic endpoints. The data presented in Tables 2 and 3 were compiled from all of the available LAL-urine assay literature, excluding only nonchromogenic spectrophotometric data[47,48] as these methodologies relied on changes in optical density over a specified time frame vs. gel-clot endpoints. Overall detection rates for nonchromogenic and chromogenic endpoint assays were calculated to be 99.1 and 93.7%, respectively. False-positive and false-negative rates for chromogenic and nonchromogenic endpoints were negligible (Table 2). The exceptionally low occurrences of reported false-positive and false-negative tests are reflected in both high positive and negative predictive

TABLE 4
Detection of 10^4 to 10^5 GNB[a]/ml by the LAL Test

Test endpoint	Total urines tested	Total 10^4—10^5 GNB/ml	Total LAL positive (%)
Gelation	1540	68	29 (42.6)
Spectrophotometric (nonchromogenic)	580	25	10 (25.0)
Chromogenic	1674	52	1 (1.9)

[a] GNB: Gram-negative bacilli.

values calculated for LAL/CLAL-urine assay (Table 3). These predictive values are reported without correction for prevalence rate.[58] Specificity of the LAL test for Gram-negative organisms (endotoxic component) and lack of reactivity with microorganisms lacking endotoxin, i.e., Gram-positive and fungal, is well recognized. Thus, the test is not able to detect significant bacteriurias of Gram-positive and fungal etiology.[31,34,35] Inclusion of these apparent false-negative tests into the data used for calculating positive and negative predictive values results in a predictable decrease in test sensitivity (Table 3); however, predictive values for a positive and negative test remain high as the great majority of clinically significant bacteriurias are of Gram-negative etiology in patient populations with high prevalence rates,[5,9] as well as for patient populations in which prevalence rates are lower.[59-61]

To date, studies performed describing LAL or CLAL screening of urine have cut-off points equal to or $>10^5$ GNB per milliliter for a positive test, as based on the original definition of UTI.[11] However, bacterial counts of $<10^5$ can be clinically significant when present with pyuria,[62] with baseline cut-off criteria as low as 10 to 100 CFU/ml in women who are symptomatic.[63] Although data for LAL results of 10^4 to $<10^5$ GNB per milliliter urine have been reported, the LAL test, as presently constituted, lacks sensitivity for Gram-negative counts of $<10^5$, as shown in Table 4.

As discussed previously, greater than 3000 LAL-urine tests have been reported in the literature, with a combined total of 46 documented false-positive assay (including bacteriurias of Gram-positive and fungal etiology) and only 14 false-negative assays (failure to detect urines with a cut-off criteria of $>10^5$ GNB per milliliter). Although these numbers are exceptionally low, possible reasons for their occurrence are discussed in the following sections.

A. FALSE-NEGATIVE LAL-URINE ASSAYS

Only 2 false-negative LAL-urine tests[32] using gel endpoints for detection of Gram-negative bacteriuria have been reported out of a combined total of 1450 urines reported (Table 2). In this study, testing was performed by diluting urine 1:100, with an equivalent of 500 ng/ml endotoxin considered a positive test. Nachum and Berzofsky[41] have reported that urine endotoxin concentrations range between 200 to 2150 ng/ml during episodes of Gram-negative bacteriuria; thus, the overall lack of sensitivity of the LAL preparation would have precluded detection of endotoxin concentrations at the lower end of the range shown to be present[41] in Gram-negative bacteriuric urines.

A total of 16 false-negative LAL reactions have been reported in which spectrophotometric or turbidometric endpoints were used for detection of Gram-negative bacteriuria.[47,48] These false-negative assays were probably not related to lysate sensitivity, as the use of spectrophotometric endpoints was shown to increase lysate sensitivity 60- to 250-fold;[45] however, initial urine dilutions of 1:2000[47] were necessary to avoid large numbers of false-positive assays, i.e., urine counts of $<10^5$ GNB per milliliter, and these specimens may have been diluted below the detection limits of the test.

Out of a tabulated 1674 chromogenic assays, including spectrophotometric and test tube methods, 12 false-negative reactions have been reported.[52] Reasons for these false-negative LAL tests may have been a result of performing the test at room temperature, affecting the sensitivity of the test by decreasing the rate of activation of the *Limulus* proclotting enzyme to its active form, a temperature dependent process. Testing of these urines at 37°C resulted in elimination of all false-negative assays. However, room temperature testing may be only part of the reason for the reported false-negative assays. Recent urine voiding could have the effect of decreasing the available free endotoxin necessary for optimal LAL test performance. Endotoxin concentrations per cell of *E. coli* have been shown to range from 32.5 to 10.0 fg/ml during logarithmic and stationary growth phases, respectively;[36] thus, the majority of endotoxin shown present in Gram-negative bacteriuric urine[41] would be expected to be cell-free and easily diluted as a result of voiding or drinking large quantities of liquid, falling below the sensitivity of CLAL room temperature assay.

B. FALSE-POSITIVE LAL-URINE ASSAYS

Clinical false-positive *Limulus* reactions can be defined by two criteria: (1) activation of the *Limulus* proclotting enzyme[64] by substances other than endotoxin and (2) activation of the clotting enzyme by endotoxin but in amounts falling outside of the test constraints. Substances such as peptidoglycans isolated from Gram-positive bacteria,[65] polynucleotides,[66] streptococcal pyrogenic exotoxins,[67] and cell-wall components isolated from *Listeria monocytogenes*[68] have been reported to cause nonspecific gelation of LAL; however, data refuting these claims have been published.[29,69,70] Simple polysaccharides,[71] synthetic dextrans,[72] and hemodialyzer rinses[73-75] have been shown to gel LAL, with the latter characterized as a cellulose derivative.[74] Maitra et al.[70] have suggested that β-hydroxy fatty acids be used as chemical markers[76,77] to confirm the presence of endotoxin in a substance claimed to cause nonspecific activation of the *Limulus* proclotting enzyme.

In the second category, LAL reactivity is endotoxin-mediated; however, a positive test may be considered as false. For example, endotoxin detected in the bloodstream in the absence of proven Gram-negative bacteremia would be interpreted as a false-positive test if the test criteria for a positive test is the presence of circulating GNB, even though the activating substance is clearly endotoxic in nature, having entered the bloodstream in the absence of Gram-negative bacteremia.[27,29] A striking example of this phenomenon was described by Garibaldi et al.[78] in patients with indwelling urine catheters. In their study, 17/24 (71%) catherized patients with Gram-negative bacteriuria were shown to have positive blood-LAL tests; in contrast, only 3/15 (20%) nonbacteriuric patients were found to have positive blood LAL assays. None of the positive blood-LAL assays were associated with Gram-negative bacteremia; however, endotoxic activity in the circulation confirmed previous reports of an association between circulating endotoxin-like activity and demonstrable foci of Gram-negative infection.[79,80]

False-positive urine-LAL tests have been reported for both nonchromogenic[32,35,47] and chromogenic[41,52] endpoint assays. Such false-positive tests can be divided into three categories: (1) specimens in which no growth ($<10^3$ CFU/ml) was reported but for which a positive LAL test was scored, (2) positive tests obtained from specimens in which $<10^5$ GNB per milliliter were isolated, and (3) positive LAL tests obtained from specimens in which $>10^5$ GNB per milliliter were isolated. The first category, i.e., no growth observed at a urine dilution of 1:1000 was discussed in Section II. The combined total of three such cases[35,52] was shown to be a result of documented or probable gonococcal infection (two urethritis and one endocervical), with cell-bound and/or cell-free gonococcal endotoxin contamination of the urine specimen postulated to have caused the positive LAL reaction. Nurminen et al.[56] obtained urine from patients with gonococcal urethritis, with 4/6 CLAL positive, supporting the notion that genito-urinary tract disease with *N. gonorrhoeae* may result in "false-positive" LAL-urine assays when a criteria of $>10^5$ GNB per milliliter urine is used.

TABLE 5
Data Summary: False-Positive LAL-Urine Tests

Test endpoint	Positive LAL Tests[a] <10^5 GNB[b]/ml	<or>10^5 GPB[c]/ml	<10^3 CFU/ml	Ref.
Nonchromogenic	0	0	0	31
	15	8	—	32
	4	0	2	35
	10	0	—	47
Chromogenic	1	8	0	41
	0	10	1	52
	0	0	0	56

[a] Mixed Gram-positive and Gram-negative urines not included.
[b] GNB: Gram-negative bacilli.
[c] GPB: Gram-positive bacilli.

Category 2 false-positive LAL-urine assays comprise data reported for counts of <10^5 GNB per milliliter when the cut-off point for a positive test was defined as ≥10^5 GNB per milliliter (Table 5). Overall, this category constituted the greatest number of reported false-positive tests, with a total of 29 nonchromogenic assays and only 1 for chromogenic endpoint assays (Table 5). Reasons for these false-positive assays may include (1) improper cleansing prior to voiding, resulting in contamination of the urine with cell-free endotoxin, a problem expected to be of greater import for female patients due to the Gram-negative microflora common to the female genito-urinary tract;[81,82] (2) endotoxin derived from different species and strains of urinary tract Gram-negative pathogens are not equally reactive with any given LAL preparation[36] (thus, it may be postulated that for some of these false-positive reactions, the endotoxin activated more of the *Limulus* proclotting enzyme over a given period of time in relation to the reference endotoxin used to measure the LAL sensitivity); and (3) test parameters, e.g., the greatest number of reported false-positive LAL-urine tests[32] were observed when a urine dilution of 1:100 and an incubation time of 2 h at 37°C was used. It is possible that the lengthy incubation time may have resulted in the reported false-positive reactions. A decrease in testing time might have obviated some of the reported false-positive LAL-urine tests; however, decreasing the incubation time might have had the effect of decreasing the sensitivity of the assay.

False-positive LAL tests from urine specimens containing >10^5 GPB per milliliter constitutes the second most commonly reported false-positive test. LAL reactivity with Gram-positive bacteria and/or substances isolated from Gram-positive cells[65,67] has been previously described. In one study[41] urine specimens which were shown to produce LAL false-positive reactions were chloroform extracted and retested for LAL reactivity. All extracts were shown to be LAL positive, indicating that LAL reactivity was endotoxin-mediated and not due to nonspecific activation of the *Limulus* proclotting enzyme. Second, in almost all of the reported false-positive LAL-urine nonchromogenic and chromogenic assays, multiple isolations of Gram-positive bacteria were usually considered contaminants[83] as a result of *in vivo* contamination of the urine specimen during collection. Contamination of the specimen could come from the external (vulva or perineum) and/or internal (vagina, cervix) genitalia; both sites are known to harbor a large indigenous microflora of Gram-positive and Gram-negative bacteria,[82] as is the female urethra.[81] These anatomical sites commonly contribute to the bacterial load of urine, adding to the endotoxin load shown present in culture-negative urine (<10^3 CFU/ml),[41] as shown in Table 6. These data support the contention that false-positive LAL assays are most likely due to contamination of urine during voiding, as Gram-negative cell counts of 1 to 2 × 10^5 or greater could not account for the endotoxin concentrations[41] shown in Table 6.

TABLE 6
Endotoxin Concentrations in Culture-Negative and Culture-Positive Urine Specimens

Patient urine	No. tested	Urine conc range (ng/ml)
True negative ($<10^3$ GNB[a]/ml)	10	3.5—20
True positive ($\geq 10^5$ GNB/ml)	20	200—2150
False positive ($<10^3 \geq 10^5$ CFU/ml)	9	40—175

[a] GNB: Gram-negative bacilli.

TABLE 7
Comparison of Predictive Values for Urine Screens for Significant Bacteriuria

Method	% Sensitivity	% Specificity	% Predictive value		Ref.
			Positive	Negative	
Nonautomated					
LAL[a]	86.8	98.3	92.0	97.7	—
Gram stain	97.4	70.6	38.4	99.1	22
Acridine Orange	99.4	58.6	30.6	99.8	21
LN strip	82.3	98.4	93.9	94.8	24
RUS[b]	42.0	97.0	71.0	92.0	25
FiltraCheck-UTI	96.5	79.7	49.1	99.1	26
Automated					
Bac-T-Screen	97.4	66.4	38.4	99.1	15
AMS Vitek	82.8	72.4	48.5	93.1	12
MS-2	74.9	87.1	64.6	91.7	12

[a] Average of nonchromogenic and chromogenic endpoint assays.
[b] RUS: rapid urine system, detection at the 50,000 CFU/ml level.

C. LAL-URINE TEST: COMPARISON TO OTHER BACTERIURIA SCREENING METHODS

Both automated and nonautomated rapid urine screening tests have been described for detection of urinary tract infection. Nonautomated methods include microscopy[23] (including gram-stained smears),[15,22] acridine orange stained smears,[21] LN strip,[17,24] rapid urine system (RUS),[25] and filtration.[26] Automated detection methods use photometry,[12-14] filtration,[15] and bioluminescence[18] to detect clinically significant urinary tract infection. Automated methods generally require expensive equipment and longer incubation times as compared to nonautomated methods. Sensitivity, specificity, and predictive values for some urine screens are compared to the LAL testing of urine for detection of bacteriuria (Table 7). LAL data were obtained from Table 3, representing an average of chromogenic and gel endpoint methods. As shown in Table 3, LAL test efficacy is high for UTI of Gram-negative etiology. Inclusion of UTIs caused by Gram-positive bacteria and fungi results in a predictable decrease in test sensitivity; however, specificity and predictive values for both positive and negative tests remain high, as the majority of significant bacteriurias ($>10^5$ CFU/ml) are caused by GNB.[34,38] A comparison of LAL-urine screening to other bacteriuria assays, automated and nonautomated, shows the LAL test to be considerably more specific, with the exception of the LN strip[24] and

RUS;[25] however, other studies[5,17] have indicated problems with the LN strip, with predictive values significantly less than shown in Table 7. Predictive negative values of greater than 90% are shown for all of the assays in Table 7; in contrast, the predictive value of a positive test is low and attributed to lack of specificity as these screens eliminate culture-negative specimens,[13] but fail to differentiate specimens containing urinary tract contaminants with counts of $>10^5$ CFU/ml from those with urinary tract pathogens. The LAL test, on the other hand, screens for bacteriuria by measuring urine endotoxin content as an index for significant numbers of Gram-negative urinary tract pathogens, thus eliminating large numbers of false-positive tests, i.e., lack of specificity, common to many urine screening methods. Since all of the assays described in Table 7 rely on the presence of large numbers of microorganisms for a positive test, no distinction can be made between probable urinary tract pathogens and contaminants. Urinary tract contaminants[83] are most frequently Gram-positive bacteria present in mixed culture, i.e., α-hemolytic streptococci, diphtheroids, lactobacilli, and coagulase-negative staphylococci. The LAL test is unable to react with these microorganisms or their metabolites, thus specificity remains high and the predictive value of a positive test takes on considerable import.

IV. CONCLUSIONS AND FUTURE TRENDS

In most clinical microbiology laboratories, urine specimens sent for culture and sensitivity are among the most numerous received. Urine screening assays of various kinds have been made available to hospital laboratories, clinics, and physician office practices[20] (for reviews, see Pezzlo[5,14]). Ultimately, the success of these screening tests will depend on a number of interrelated factors, including ease of performance, interpretability, speed, reliability, and cost. LAL-urine testing for detection of Gram-negative bacteriuria easily meets these expectations, irrespective of the format used for assay. Use of chromogenic substrates has further simplified the assay as inadvertent dissolution of the gel is obviated, thereby increasing the ease with which the assay is performed and providing an easily read and interpreted endpoint which can be readily automated.[56] The LAL testing of urine offers speed, with reaction times as short as 5 to 6 min[41,52] for chromogenic assay, comparing favorably with other nonautomated urine screens.[5] To date, no commercial test kits are available, making cost comparisons to kits currently on the market impossible; however, it would be expected that such a LAL-urine screen would be competitive on a cost per test basis.

The reliability of the LAL-urine test, i.e., high sensitivity and specificity and high positive and negative predictive values, has been shown in Section III. The inability of the LAL test to detect significant bacteriuria of Gram-positive and fungal etiologies diminish the usefulness of the assay, even though the majority of bacteriurias are of Gram-negative etiology.[5,9] At present, no equally rapid or simple tests are available for detection of Gram-positive bacteriuria. A second problem with LAL testing of urine is the detection of Gram-negative bacterial counts of $<10^5$, an exercise which at best is problematical in terms of outcome (see Table 4). Studies to date have not addressed this problem. In light of recent investigations in which bacterial counts of $<10^5$ and ranging as low as 10 to 100/ml urine[62,63] may signify UTI in patients who are symptomatic, detection of bacteriurias with lower bacterial counts cannot be ignored in patient populations with high prevalence rates. Although LAL detection of $<10^5$ GNB per milliliter is feasible, assay reliability could pose a problem, as culture-negative urine has been shown to contain background levels of endotoxin.[41]

Given the above, is there a place for the LAL testing of urine for detection of significant bacteriuria? The answer may lie in consideration of carefully selected patient populations for whom a cut-off criteria of $\geq 10^5$ would be applicable, based on known prevalence rates[15,63] of 10% or lower. Several large patient populations meeting these criteria are known, e.g., bacteriuria of pregnancy,[84-86] the elderly,[61] males, hospitalized patients and single females,[8] school age children,[6] and infants.[3] For those patient populations with low prevalence rates and for the vast majority of patients whose UTIs are caused by GNB the LAL-urine test may provide a useful screening method for rapid detection of $\geq 10^5$ GNB per milliliter, especially when one considers the high LAL positive and negative predictive values of the test.

REFERENCES

1. **Shlaes, D. M.,** From the simple to the complicated, *Diagnosis*, October, 73, 1982.
2. **Kass, E. H.,** Bacteriuria and the diagnosis of infections of the urinary tract, *Arch. Intern. Med.*, 100, 709, 1957.
3. **Kunin, C. M.,** Epidemiology and natural history of urinary tract infection in school age children, *Pediatr. Clin. North Am.*, 18, 509, 1971.
4. **Neu, H. C.,** Urinary tract infections in the 1980s, *Semin. Urol.*, 1(2), 130, 1983.
5. **Pezzlo, M.,** Detection of urinary tract infection by rapid methods, *Clin. Microbiol. Rev.*, 1, 268, 1988.
6. **Downs, A. W. and Cleland, V. S.,** Bacteriuria and urinary tract infection in infancy and childhood: a review, *Nurs. Res.*, 20, 131, 1971.
7. **Yoshikawa, T. T.,** Unique aspects of urinary tract infection in the geriatric population, *Gerentology*, 30, 339, 1984.
8. **Mond, N. C., Percival, A., Williams, J. D., and Brumfitt, W.,** Presentation, diagnosis, and treatment of urinary-tract infections in general practice, *Lancet*, 1, 514, 1965.
9. **Blazevic, D. J., Stemper, J. E., and Matsen, J. M.,** Organisms encountered in urine cultures over a 10-year period, *Appl. Microbiol.*, 23, 421, 1972.
10. **Barry, A. L., Smith, P. B., and Turck, M.,** Laboratory diagnosis of urinary tract infections, in *Cumitech*, Vol. 2, Gavin, T. L., Ed., American Society of Microbiology, Washington, D.C., 1975.
11. **Kass, E. H.,** Asymptomatic infections of the urinary tract, *Trans. Assoc. Am. Phys.*, 69, 56, 1956.
12. **Pezzlo, M. T., Tan, G. L., Peterson, E. M., and de la Maza, L. M.,** Screening of urine cultures by three automated systems, *J. Clin. Microbiol.*, 15, 468, 1982.
13. **Pezzlo, M.,** Detection of bacteriuria by automated methods, *Lab. Med.*, 15, 539, 1984.
14. **Pezzlo, M. T.,** Automated methods for detection of bacteriuria, *Am. J. Med.*, 75(Suppl. 13), 71, 1983.
15. **Davis, J. R., Stager, C. E., and Araj, G. F.,** Clinical laboratory evaluation of a bacteriuria detection device for urine screening, *Am. J. Clin. Pathol.*, 81, 48, 1984.
16. **Cashman, J. S., Ahlin, P. A. and Matsen, J. M.,** Evaluation of a rapid bacteriuria screening instrument, *Diagn. Microbiol. Infect. Dis.*, 2, 11, 1984.
17. **Malloy, P. J., Schreckenberger, P. C., and Lebeau, L. J.,** Evaluation of five urine screen techniques, *Abstr. Annu. Mtg. Am. Soc. Microbiol.*, C–186, 267, 1984.
18. **Welch, W. D., Thompson, L., Layman, M., and Southern, P. M., Jr.,** Evaluation of two bioluminescence-measuring instruments, the Turner design and Lumac systems, for the rapid screening of urine specimens, *J. Clin. Microbiol.*, 20, 1165, 1984.
19. **Ellner, P. D. and Papachristos, T.,** Detection of bacteriuria by dip-slide. Routine use in a large general hospital, *Am. J. Clin. Pathol.*, 63, 516, 1975.
20. **Fowler, J. E., Jr.,** Office bacteriology: techniques and interpretations, *Sem. in Urolol.*, 1, 97, 1983.
21. **Hoff, R. G., Newman, D. E., and Staneck, J. L.,** Bacteriuria screening by use of acridine orange-stained smears, *J. Clin. Microbiol.*, 21, 513, 1985.
22. **Washington, J. A., II, White, C. M., Laganiere, M., and Smith, L. H.,** Detection of significant bacteriuria by microscopic examination of urine, *Lab. Med.*, 12, 294, 1981.
23. **Kunin, C. M.,** The quantitative significance of bacteria visualized in the unstained urinary sediment, *N. Engl. J. Med.*, 265, 589, 1961.
24. **Smally, D. L. and Dittman, A. R.,** Use of leucocyte esterase-nitrate activity as predictive assays of significant bacteriuria, *J. Clin. Microbiol.*, 18, 1256, 1983.
25. **Vail, S., Gruninger, R. P., Hanson, K., and Klicker, R. E.,** Clinical evaluation of LYFO Kwik (R) rapid urine system (RUS), *Abstr. Annu. Mtg. Am. Soc. Microbiol.*, C–154, 357, 1988.
26. **Longaria, C. C. and Gonzalez, G. A.,** FiltraCheck-UTI, a rapid disposable system for detection of bacteriuria, *J. Clin. Microbiol.*, 25, 926, 1987.
27. **Nachum, R. and Neely, M.,** Clinical diagnostic usefulness of the Limulus amoebocyte lysate assay, *Lab. Med.*, 13, 112, 1982.
28. **Prior, R. B.,** Bacterial endotoxin detection by the *Limulus* amoebocyte lysate test, in *Antigen Detection to Diagnose Bacterial Infections*, Vol. 2, Kohler, R. B., Ed., CRC Press, Boca Raton, FL, 1986, chap. 30.

29. **Siegel, S. E. and Nachum, R.,** Use of the Limulus lysate assay (LAL) for the detection and quantitation of endotoxin, in *Perspectives in Toxicology*, Bernheimer, A. W., Ed., John Wiley & Sons, New York, 1977, 61.
30. **Nachum, R., Lipsey, A. L., and Siegel, S. E.,** Rapid detection of gram-negative bacterial meningitis by the *Limulus* test, *N. Engl. J. Med.*, 289, 931, 1973.
31. **Jorgensen, J. H., Carvajal, H. F., Chipps, B. E., and Smith, R. F.,** Rapid detection of Gram-negative bacteriuria by use of the *Limulus* endotoxin assay, *Appl. Microbiol.*, 26, 38, 1973.
32. **Jorgensen, J. H. and Jones, P. M.,** Comparitive evaluation of the Limulus assay and the direct Gram stain for detection of significant bacteriuria, *Am. J. Clin. Pathol.*, 63, 142, 1975.
33. **Cooper, J. F., Levin, J., and Wagner, H. N., Jr.,** Quantitative comparison of *in vitro* and *in vivo* methods for the detection of endotoxin, *J. Lab. Clin. Med.*, 78, 138, 1971.
34. **Levin, J. and Bang, F. B.,** The role of endotoxin in the extracellular coagulation of Limulus blood, *Bull. Johns Hopkins Hosp.*, 115, 265, 1964.
35. **Nachum, R. and Shanbrom, E.,** Rapid detection of Gram-negative bacteriuria by *Limulus* amoebocyte lysate assay, *J. Clin. Microbiol.*, 13, 158, 1981.
36. **Sullivan, J. D., Jr., Valois, F. W., and Watson, S. W.,** Endotoxin: the Limulus amoebocyte lysate system, in *Mechanisms in Bacterial Toxicology*, Bernheimer, A. W., Ed., John Wiley & Sons, New York, 1976, 217.
37. **Prior, R. B. and Spagna, V. A.,** Response of several *Limulus* amoebocyte lysates to native endotoxin present in gonococcal and nongonococcal urethral exudates from human males, *J. Clin. Microbiol.*, 13, 167, 1981.
38. **Krause-Hooyman, L. and Komorowski, R.,** Detection of Gram-negative bacteria in urine by an instrument-free LAL test, *Abstr. Annu. Mtg. Am. Soc. Microbiol.*, C–304, 374, 1987.
39. **Nachum, R.,** Usefulness of LAL assay for rapid detection of Gram-negative bacteriuria, in *Detection of Bacterial Endotoxin with the Limulus Amebocyte Lysate Test,* Watson, S. W., Levin, J., and Novitsky, T. J., Eds., Alan R. Liss, New York, 1987, 431.
40. **Miller, M. A., Prior, R. B., Horvath, F. and Hjelle, J. T.,** Comparison of the *Limulus* amoebocyte lysate (LAL) assay and urine cultures for detection of renal bacteriuria in polycystic kidney disease (PKD) patients, *Abstr. Annu. Mtg. Am. Soc. Microbiol.*, C–156, 358, 1988.
41. **Nachum, R. and Berzofsky, R. N.,** Chromogenic *Limulus* amoebocyte lysate assay for rapid detection of Gram-negative bacteriuria, *J. Clin. Microbiol.*, 21, 759, 1985.
42. **Levin, J. and Bang, F. B.,** Clottable protein in *Limulus*: its localization and kinetics of its coagulation by endotoxin, *Thromb. Diath. Haemorrh.*, 19, 186, 1968.
43. **Teller, J. D. and Kelley, K. M.,** A turbidometric *Limulus* amebocyte assay for the quantitative determination of Gram-negative bacterial endotoxin, in *Biomedical Application of the Horseshoe Crab (Limulidae)*, Cohen, E., Ed., Alan R. Liss, New York, 1979, 423.
44. **Novitsky, T. J., Remillard, J. F., and Loy, N.,** Design criteria and evaluation of the LAL-4000 for kinetic turbidometric assay, in *Detection of Bacterial Endotoxins with the Limulus Amebocyte Lysate Test*, Watson, S. W., Levin, J., and Novitsky, T. J., Eds., Alan R. Liss, New York, 1987, 189.
45. **Jorgensen, J. H. and Alexander, G. A.,** Automation of the *Limulus* amoebocyte lysate test by using the Abbott MS-2 microbiology system, *Appl. Environ. Microbiol.*, 41, 1316, 1981.
46. **Remillard, J. F., Gould, M. C., Roslansky, P. F., and Novitsky, T. J.,** in *Detection of Bacterial Endotoxins with the Limulus Amebocyte Lysate Test*, Watson, S. W., Levin, J., and Novitsky, T. J., Eds., Alan R. Liss, New York, 1987, 197.
47. **Jorgensen, J. H. and Alexander, G. A.,** Rapid detection of significant bacteriuria by use of an automated *Limulus* amoebocyte lysate assay, *J. Clin. Microbiol.*, 16, 587, 1982.
48. **Jorgensen, J. H., Redding, J. S., Reichler, A. S., and Spencer, H. J.,** Spectrophotometric Limulus amoebocyte lysate assay for rapid detection of Gram-negative bacterial meningitis of significant bacteriuria, *Abstracts, Interscience Conference on Antimicrobials and Chemotherapy*, No. 118, American Society for Microbiology, Washington, D.C., 1984, 110.
49. **Iwanaga, S. Morita, T., Harada, T., Nakamura, S., Niwa, M., Takada, K., Kimura, T., and Sakakibara, S.,** Chromogenic substrates for horseshoe crab clotting enzyme. Its application for the assay of bacterial endotoxins, *Hemostasis*, 7, 183, 1978.
50. **Jorgensen, J. H. and Smith, R. F.,** Measurement of bound and free endotoxin by the Limulus assay, *Proc. Soc. Exp. Biol. Med.*, 146, 1024, 1974.
51. **Levin, J., Tomasulo, P. A., and Oser, R. S.,** Detection of endotoxin in human blood and demonstration of an inhibitor, *J. Lab. Clin. Med.*, 75, 903, 1970.
52. **Nachum, R., Arce, J. J., and Berzofsky, R. N.,** Gram-negative bacteriuria of pregnancy: rapid detection by a chromogenic *Limulus* amoebocyte lysate assay, *Obstet. Gynecol.*, 68, 215, 1986.
53. **Shands, J. W., Jr.,** The physical structure of bacterial lipopolysaccharides, in *Microbial Toxins*, Vol. 5, Kadis, S., Weinbaum, G., and Ajl, S. J., Eds., Academic Press, New York, 1971, 127.
54. **Yin, E. T., Galanos, C., Kinsky, S., Bradshaw, R. A., Wessler, S., Luderitz, O., and Sarmiento, M. A.,** Picogram-sensitive assay for endotoxin: gelation of *Limulus polyphemus* blood cell lysate induced by purified lipopolysaccharides and lipid A from gram-negative bacteria, *Biochim. Biophys. Acta*, 261, 284, 1972.

55. **Vecchio, T. J.,** Predictive value of a single diagnostic test in unselected populations, *N. Engl. J. Med.*, 274, 1171, 1966.
56. **Nurminen, M., Karvonen, M., and Siitonen, A.,** Detection of Gram-negative bacteria in urine by the chromogenic Limulus assay, *Eur. J. Clin. Microbiol. Infect. Dis.*, 7, 529, 1988.
57. **Luke, K. E., Michaels, A. A., and Proud, L. J.,** Rapid detection of Gram-negative bacteria in urine using a *Limulus* amebocyte lysate dip and read test, *Abstr. Annu Mtg. Am. Soc. Microbiol.*, C-157, 358, 1988.
58. **Ransohoff, D. F. and Feinstein, A. R.,** Problems of spectrum and bias in evaluating the efficacy of diagnostic tests, *N. Engl. J. Med.*, 299, 926, 1978.
59. **Gordon, S. F.,** Asymptomatic bacteriuria of pregnancy, *Clin. Med.*, 79, 22, 1972.
60. **Monif, R. G.,** *Infectious Disease. Obstetrics*, 3rd ed., Infectious Disease, Inc., Gainsville, FL, 1982, 193.
61. **Dontas, A. S., Kasvike-Charvati, P., Papaanayiotou, P. C., and Marketos, S. G.,** Bacteriuria and survival in old age, *N. Engl. J. Med.*, 304, 939, 1981.
62. **Stamm, W. E., Counts, G. W., Running, K. R., Fihn, S., Turck, M., and Holmes, K. K.,** Diagnosis of coliform infection in acutely dysuric women, *N. Engl. J. Med.*, 307, 463, 1982.
63. **Barlett, R. C. and Galen, R. S.,** Predictive value of urine culture, *Am. J. Clin. Pathol.*, 79, 756, 1983.
64. **Sullivan, J. D., Jr. and Watson, S. W.,** Purification and properties of the clotting enzyme from Limulus lysate, *Biochem. Biophys. Res. Commun.*, 66, 848, 1975.
65. **Wildfeur, A., Heymer, B., Schleifer, K. H., and Haferkamp, O.,** Investigations on the specificity of the Limulus test for the detection of endotoxin, *Appl. Microbiol.*, 28, 867, 1974.
66. **Elin, R. J. and Wolff, S. M.,** Non-specificity of the Limulus amoebocyte lysate test: positive reactions with polynucleotides and proteins, *J. Infect. Dis.*, 128, 349, 1973.
67. **Brunson, K. W. and Watson, D. W.,** Limulus amoebocyte reaction with streptococcal pyrogenic exotoxin, *Infect. Immun.*, 14, 1256, 1976.
68. **Wexler, H. and Oppenheim, J. D.,** Isolation, characterization, and biological properties of an endotoxin-like material from the Gram-positive organism *Listeria monocytogenes*, *Infect. Immun.*, 23, 845, 1979.
69. **Yin, E. T.,** Endotoxin, thrombin, and the *Limulus* amebocyte lysate test, *J. Lab. Clin. Med.*, 86, 430, 1975.
70. **Maitra, S. K., Nachum, R. and Pearson, F. C.,** Establishment of beta-hydroxy fatty acids as chemical marker molecules for bacterial endotoxin by gas chromatography-mass spectrometry, *Appl. Environ. Microbiol.*, 52, 510, 1986.
71. **Mikami, T., Nagase, T., Matsumoto, T., Suzuke, S., and Suzuki, M.,** Gelation of *Limulus* amebocyte lysate by simple polysaccharides, *Microbiol. Immunol.*, 26, 403, 1982.
72. **Suzuki, M., Mikami, T., Matsumoto, T., and Suzuki, S.,** Gelation of *Limulus* lysate by synthetic dextran derivatives, *Microbiol. Immunol.*, 21, 419, 1977.
73. **Carson, L. A. and Petersen, N. J.,** LAL-reactive material associated with hemodialysis membranes, *Prog. Clin. Biol. Res.*, 93, 217, 1982.
74. **Pearson, F. C., Bohon, J., Lee, W., Bruszer, G., Sagona, M., Jakubowski, G., Dawe, R., Morrison, D. and Dinarello, C.,** Characterization of *Limulus* amoebocyte lysate-reactive material from hollow-fiber dialyzers, *Appl. Environ. Microbiol.*, 48, 1189, 1984.
75. **Petersen, N. J., Carson, L. A. and Favero, M. S.,** Bacterial endotoxin in new and used hemodialyzers. A potential cause of endotoxemia, *Trans. Am. Soc. Artificial Organs*, 27, 155, 1981.
76. **Luderitz, O., Galanos, C., Lehmann, V., Nurminen, M., Rietschel, E. T., Rosenfelder, G., Simon, M. and Westphal, O.,** Lipid A: chemical structure and biological activity, in *Bacterial Polysaccharides*, Kass, E. H. and Wolff, S. M., Eds., University of Chicago Press, Chicago, 1973, 9.
77. **Rietschel, E. T., Gottert, H., Luderitz, O. and Westphal, O.,** Nature and linkages of the fatty acids present in the lipid A component of Salmonella polysaccharides, *Eur. J. Biochem.*, 28, 166, 1972.
78. **Garibaldi, R. A., Allman, G. W., Larsen, D. H., Smith, C. B. and Burke, J. P.,** Detection of endotoxemia by the Limulus test in patients with indwelling urinary catheters, *J. Infect. Dis.*, 128, 551, 1973.
79. **Levin, J., Poore, T. E., Young, N. S., Margolis, S., Zauber, N. P., Townes, A. S. and Bell, W. R.,** Gram-negative sepsis: detection of endotoxemia with the Limulus test. With studies of associated changes in blood coagulation, serum lipids, and complement, *Ann. Intern. Med.*, 76, 1, 1972.
80. **Levin, J., Poore, T. E., Zauber, N. P. and Oser, R. S.,** Detection of endotoxin in the blood of patients with sepsis due to Gram-negative bacteria, *N. Engl. J. Med.*, 283, 1313, 1970.
81. **Cox, C. E.,** The urethra and its relationship to urinary tract infection: the flora of the normal female urethra, *South. Med. J.*, 59, 621, 1966.
82. **Lindner, J. G. E. M., Planatema, F. H. F. and Hoogkamp-Korstanje, J. A. A.,** Quantitative studies of the vaginal flora of healthy women and of obstetric and gynaecological patients, *J. Med. Microbiol.*, 11, 233, 1978.
83. **Marrie, T. J., Harding, K. M. and Ronald, A. R.,** Anaerobic and aerobic urethral flora in healthy females, *J. Clin. Microbiol.*, 8, 67, 1978.
84. **Little, P. J.,** The incidence of urinary tract infection in 5000 pregnant women, *Lancet*, 2, 925, 1966.
85. **Marchant, D. J.,** Urinary tract infections in pregnancy, *Clin. Obstet. Gynecol.*, 21, 921, 1978.
86. **Kass, E. H.,** Horatio at the orifice: the significance of bacteriuria, *J. Infect. Dis.*, 138, 546, 1978.

Chapter 8

EVALUATION OF GONORRHEA

Richard B. Prior

TABLE OF CONTENTS

I. Introduction 98

II. Urethritis 98

A. Epidemiology of Urethritis 98
B. Demography and Clinical Characteristics of Urethritis 99
C. The Gram Stain in Evaluation of Urethritis 99
D. Noncultural Methods to Detect Gonorrhea 100

III. Presumptive Diagnosis of Gonococcal Urethritis in Males Using the LAL Test 100
A. The Hypothesis 100
B. Determination of Test Parameters 101
C. Development of Test Kit 102
D. Clinical Results 102
E. New Developments 104

IV. Evaluation of Gonococcal Cervicitis 105

V. Evaluation of Sexual Contacts and Disseminated Gonorrhea 106

VI. Summary 107

References 107

I. INTRODUCTION

Sexually transmitted diseases (STDs) are epidemic throughout the world. Traditionally, gonorrhea and syphilis were thought of as *the* "venereal diseases"; however, more STDs have recently been recognized. Currently, there are over 20 different types that are transmitted by sexual contact.[1] In the U.S. alone, there are over 3 million cases of chlamydia (caused by the bacterium *Chlamydia trachomatis*), 200,000 to 500,000 new cases of genital herpes, and at least 1 million cases each of pelvic inflammatory disease (PID) (almost always caused by chlamydial and/or gonococcal infection), trichomonal infection, and genital warts (caused by human papillomavirus).[2] Other less known STDs, such as chancroid and bacterial vaginosis, are also on the rise.[2] Syphilis is the third most frequently reported communicable disease in the U.S. exceeded only by varicella and gonorrhea.[3] The most frequently reported disease is gonorrhea, with a median for 1983 through 1987 of 883,826 civilian cases per year.[4] In 1988, 719,536 civilian cases of gonorrhea were reported to the Centers for Disease Control (CDC), compared to 1,004,029 cases reported in 1980, showing a decrease in this disease.[5] However, gonorrhea is still occurring in epidemic proportions. Finally, acquired immunodeficiency syndrome (AIDS) has been increasing, with 30,847 cases reported in the U.S. during 1988.[4] The majority of AIDS cases are transmitted by sexual contact. STDs are so prevalent today that it is estimated that a majority of graduating physicians are going into specialties where they may expect 1 in 20 patients (5% of their practice) to have an STD.[2]

Control measures for STDs are directed toward education, effective diagnosis and treatment, and appropriate follow-up and contact tracing. However, human and financial resources are strained, and newer methods and approaches in the evaluation of STDs are always being sought in an effort to reduce these serious health problems.

The use of the *Limulus* amoebocyte lysate (LAL) test in evaluating gonococcal infections has been studied extensively in both males and females. In this chapter, the major emphasis is on the use of the LAL test in evaluating male urethritis, since most of the work has been done in this clinical syndrome. A review of the epidemiology, demography, and clinical characteristics of urethritis is presented, followed by a discussion of the Gram stain in the evaluation of urethritis. The use of the LAL test in evaluating urethritis and gonococcal cervicitis is also discussed, as is the use of the test in evaluating sexual contacts and patients with disseminated gonococcal infection.

II. URETHRITIS

A. EPIDEMIOLOGY OF URETHRITIS

Among the many diseases transmitted by sexual contact, gonorrhea is the most commonly reported, with approximately 800,000 cases reported to the CDC annually in the U.S. Of these cases, the majority occurs in males (gonococcal urethritis [GU]), and it is estimated that cases equaling four to five times this number are not reported but are seen by physicians in the private sector.[6] The etiologic agent of gonorrhea is the Gram-negative diplococcus *Neisseria gonorrhoeae*.

Nongonococcal urethritis (NGU) includes all cases of infectious urethritis caused by known or suspected pathogens other than *N. gonorrhoeae*. Thus, the diagnosis of NGU is made by ruling out the presence of gonorrhea. Two established pathogens of NGU are *C. trachomatis* and *Ureaplasma urealyticum*, with *C. trachomatis* isolated from approximately 40% of men with NGU.[7] However, in more than 50% of NGU cases, the infectious agents are not identifiable due to a limited understanding of the complex flora of the normal urethra. The myriad of different aerobic and anaerobic microorganisms isolated from cases of NGU and normal controls has made the distinction between pathogenic and commensal microorganisms an incredibly difficult task.[8] The epidemiology of NGU is a matter of speculation and defies precise definition

TABLE 1
Clinical Breakdown of Urethral Exudates from 967 Men with Culture-Positive Gonococcal and Nongonococcal Urethritis

		Percent urethral discharge				
Diagnosis	No. evaluated	Profuse[a]	Moderate[b]	Scant[c]	Purulent	Nonpurulent
Gonococcal urethritis	603	39	40	21	67	33
Nongonococcal urethritis	364	12	44	44	23	77

[a] Spontaneous discharge visible without penile massage.
[b] Discharge not obvious, but visible upon compression of the glans.
[c] Discharge obtained only after urethral massage.

Reprinted from Spagna, V. A., Urethritis syndromes, in *Sexually Transmitted Diseases: A Clinical Syndrome Approach*, Spagna, V. A. and Prior, R. B., Eds., Marcel Dekker, New York, 1985, p. 75. Courtesy of Marcel Dekker, New York.

because of (1) the lack of definite diagnostic criteria, (2) ignorance about disease incidence and prevalence secondary to the worldwide absence of systematic recordkeeping, and (3) the paucity of information about risk factors.[9] It is now estimated that NGU is nearly twice as frequent as gonorrhea in males, and family physicians can expect cases of NGU to comprise 50 to 70% of males with urethritis.[9,10]

B. DEMOGRAPHY AND CLINICAL CHARACTERISTICS OF URETHRITIS

Patients with GU or NGU are similar with respect to age, marital status, and number of different sexual contacts. Persons with NGU are more likely to be educated and to have an established relationship with a private physician; however, no specific demographic features are helpful in predicting the presence or absence of GU.[11] Several clinical characteristics differentiating GU from NGU have been formulated. Simultaneous discharge and dysuria, a spontaneous purulent exudate, and an abrupt, severe onset suggest gonorrhea. The discharge associated with NGU is typically white, mucoid, scant, and gradual in onset. The observation of a discharge only in the morning before voiding is reported by 29% of patients with NGU.[12] While it is apparent that the classic signs and symptoms of NGU and GU differ, there is sufficient overlap (as shown in Table 1) that an accurate diagnosis can be achieved only by utilizing laboratory tests.[13-16] The optimal performance of clinical diagnosis based solely on history and the nature of the urethral discharge is 73%.[17]

There has been a trend toward treating all urethritis cases with broad spectrum, multidose antibiotic regimens as a compromise to a specific diagnosis. While this approach is sound in theory, it does require strict patient compliance and does not permit a sound orientation toward the disposition of sexual contacts. One study compared the treatment results achieved for GU by 5 d courses of tetracycline HCl and doxycycline.[18] The compliance rates for tetracycline HCl and doxycycline were 62 and 78.5%, respectively. Single-dose therapy is effective in treating GU, and efforts should be made to diagnose urethritis accurately and expediently.

C. THE GRAM STAIN IN EVALUATION OF URETHRITIS

Currently, the Gram stain is the only acceptable immediate diagnostic aid for the presumptive diagnosis of urethritis in males.[15] When read by *experienced* microscopists, Gram-stained smears of urethral exudate from males with symptomatic urethritis are highly sensitive (93 to

99%) and highly specific (98%).[15,19,20] However, experience of the microscopist is the most important factor in obtaining accurate results. In a recent study, Gram stain sensitivity of urethral smears varied from 79 to 96% and specificity varied from 52 to 97%, depending on the microscopist reading the smear.[21] The microscopists who obtained the highest sensitivity and specificity were those working in an STD clinic where they had been reading more than 200 smears per week for several years. Unfortunately, such expertise is usually restricted to an STD clinic and may not be available to physicians practicing in the private sector. A survey conducted at the 1981 annual meeting of the American Academy of Family Physicians indicated that 31% of the private physicians did not do Gram stains in their offices; of those who did, the majority felt that the Gram stain was less than 80% accurate in their office setting.[22] This was also shown in a recent study by Landis et al.,[23] who reported a relatively low sensitivity (66%) and negative predictive value (63%) with the Gram stain and concluded that the test was of limited value in diagnosis and therapeutic decision-making when the patient was first seen. Thus, cultures for *N. gonorrhoeae* remain the absolute standard for the diagnosis of GU. However, culture results are obtained only after considerable delay (usually 48 h), and the fastidious growth requirements of the gonococci can adversely affect the sensitivity of culture.

D. NONCULTURAL METHODS TO DETECT GONORRHEA

An alternative to the cultural diagnosis of gonorrhea has been the use of noncultural methods such as enzyme-linked immunosorbent assays (ELISAs) and nucleic acid probes. These noncultural methods offer several advantages over the standard cultural procedures because the diagnosis is not dependent on the presence of viable bacteria and turnaround times are greatly reduced. Commercially available ELISA systems for the direct detection of both the gonococci and chlamydia in clinical samples are available (Gonozyme and Chlamydiazyme, Abbott Laboratories, North Chicago, IL), with reports of good correlation with culture results.[24-28] These systems do require specialized equipment, including a spectrophotometer, and while the hands-on time is only about 30 min, the total testing time is approximately 4 h.

Newer nonisotopic DNA probes and filtration ELISA tests using monoclonal antibodies are now being evaluated as quick and simple "test kit" methods to detect *N. gonorrhoeae* and *C. trachomatis* directly in urogenital samples.[29-33] Unlike the Abbott systems, these tests do not require expensive equipment and total testing time is less than 1 h. These assays look very promising but further studies are needed to determine their clinical utility.

III. PRESUMPTIVE DIAGNOSIS OF GONOCOCCAL URETHRITIS IN MALES USING THE LAL TEST

The current epidemic of GU and NGU will continue unabated until accurate, on-the-spot evaluation, with proper treatment of both patients and sexual contacts, becomes widespread in the private sector. Numerous studies have shown the potential usefulness of the LAL test for the evaluation of males with exudative urethritis.[34-42]

A. THE HYPOTHESIS

The LAL test was initially studied as a diagnostic aid in the evaluation of exudative urethritis in males based on the hypothesis that gonococcal urethral exudates would contain high concentrations of endotoxin, while nongonococcal urethral exudates would contain lesser amounts. The gonococci (*N. gonorrhoeae*) contain endotoxin in their cell walls and also shed endotoxin into the surrounding environment during the growth phase and upon autolysis (cell death). Conversely, the known and proposed etiologic agents of NGU elaborate little or no endotoxin. Therefore, by properly diluting urethral exudates, endotoxin activity associated with NGU can be easily eliminated while retaining a high degree of sensitivity to the endotoxin from *N. gonorrhoeae*.

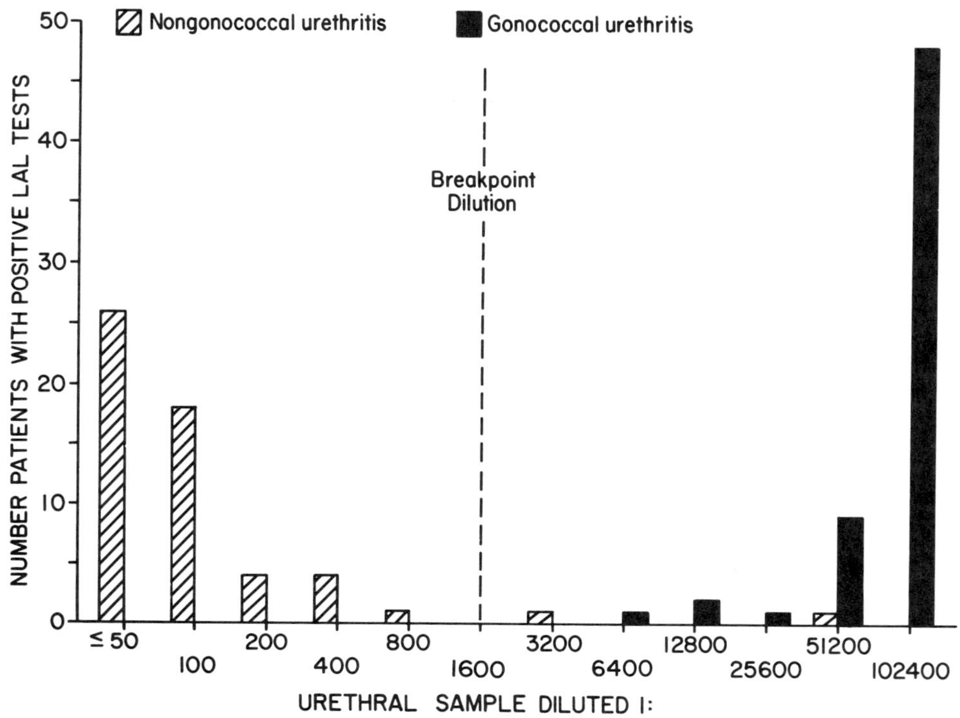

FIGURE 1. Relationship between dilution of urethral sample and positive LAL results for 116 male patients with culture-proven gonococcal and nongonococcal urethritis. Optimum dilution for maximum sensitivity and specificity is 1:1600 for the sensitivity of lysate used.

B. DETERMINATION OF TEST PARAMETERS

Limulus lysate, volume of collected exudate, and dilution of the clinical sample are variables that would affect LAL test results. By controlling lysate sensitivity and the volume of exudate collected, the optimum dilution of sample for maximum sensitivity and specificity was determined in clinical trials where quantitative studies of the endotoxin content in urethral exudates were performed.[37,43] Following collection of about 25 µl of urethral exudate using a tuberculin syringe without a needle and employing gentle aspiration, samples were diluted in pyrogen-free water from 1:50 to 1:102,400 in 2-fold amounts. Each dilution was then tested for endotoxin activity using a lysate with a minimum sensitivity of 0.0625 ng/ml, U.S. Standard Endotoxin Lot EC-2. Results of these quantitative studies are shown in Figure 1. The optimum breakpoint dilution was 1:1600, which represents 100 ng of endotoxin per milliliter of exudate (1600×0.0625 ng/ml = 100 ng/ml). Thus, urethral exudates which contain 100 ng or more of endotoxin per milliliter will correlate with positive cultures for *N. gonorrhoeae* in over 98% of cases.[38] The exact dilution needed for lysates of different sensitivities can be computed using the 100 ng/ml breakpoint value and the equation $D = k/MS$ where D is the dilution of clinical sample (±1 2-fold amount), assuming a collection of 25 µl of exudate; k is the constant endotoxin concentration of 100 ng/ml representing the minimum amount of endotoxin in urethral exudate needed to predict gonorrhea; and MS is the minimum sensitivity of lysate in nanogram per milliliter (EC-2 endotoxin standard).[38] For example, given a lysate sensitivity of 0.25 ng/ml, the proper dilution needed would be 1:400, which corresponds to a diluent volume of 10 ml. These values can be converted into endotoxin units (EU) assuming 5 EU = 1 ng.

Clinical data also showed that the standard incubation time of 60 min could be reduced to 30 min with no statistically significant differences in test results.[38] Additionally, the data in Figure 1 show that collection of a precise amount of exudate is not critical. A 50% change in volume of exudate collected will not significantly change the sensitivity of the test.[27,44]

C. DEVELOPMENT OF TEST KIT

Based on the information above, a test kit for the rapid evaluation of exudative urethritis in males was developed (Gonoscreen), patented,[45] and approved by the U.S. Food and Drug Administration (FDA) in 1983 for the presumptive diagnosis of gonococcal urethritis. The test kit contained the necessary materials for the "in office" evaluation of exudative urethritis. The kit included (1) a 1-ml syringe used for specimen collection, (2) a sealed dilution ampule containing 10 ml of pyrogen-free water, (3), a single-test LAL vial containing lyophilized lysate with a sensitivity of 0.625 to 2.5 EU/ml, and (4) a plastic needle used to break the dilution ampule seal and also transfer the diluted sample to the LAL vial. The dilution and transfer steps were done in a closed system to prevent contamination of the sample and biohazards to the user. The basic procedure for using the LAL test kit for evaluating exudative urethritis is shown in Figure 2.

D. CLINICAL RESULTS

From 1979 to 1983, the LAL test was used to evaluate 1812 male patients with exudative urethritis seen at the City Health Department STD clinic, Columbus, OH. The results, comparing the LAL test to Gram stain, are shown in Table 2. Sensitivity and specificity of the LAL test were 99.3 and 96.8%, respectively, with an overall accuracy of predicting cultures for *N. gonorrhoeae* of 98.3%. The results of clinical studies involving 2283 cases of exudative urethritis evaluated by the LAL test and reported in the literature are summarized in Table 3. The overall ability of the LAL test to predict culture results for *N. gonorrhoeae* ranged from 94 to 100%. In addition, the Gonoscreen test kit was evaluated in five separate clinical studies throughout the U.S. The comparative results of Gram stain and the test kit for 1100 male patients with exudative urethritis are shown in Table 4. The sensitivity and specificity were 97.5 and 95.1%, respectively, with an overall accuracy of 96.9%. The results were not statistically different from Gram stain smears which were read by experienced microscopists or from cultures processed in the STD clinics.

The positive predictive values of the LAL test for various theoretical prevalence rates of GU computed according to the method described by Vecchio[46] are shown in Table 5. In private practice settings, the relative prevalence of gonorrhea among men with acute urethritis would be about 30%.[47] The predictive value of a positive LAL test at this frequency of gonorrhea would be 90%. Thus, a positive test justifies a diagnosis of GU, allowing appropriate management of patients and their sexual contacts. Since the predictive value varies with the prevalence of GU, it is recommended that clinicians be familiar with the prevalence of GU in their patient populations when interpreting test results.[41]

Approximately 3% of the LAL test results did not correctly predict culture results. Although the reasons for these discrepancies are speculative, the possibility of culture failures, vancomycin-susceptible strains, contaminated equipment, or technical error may provide an explanation. In addition, the LAL test detects endotoxin and the viability of the gonococci has no influence on the test. Therefore, when using culture as the standard, unreported antibiotic usage by the patient would also yield a "false" positive LAL test.[43]

The advantages of the LAL test in the evaluation of exudative urethritis in men are speed, utility, simplicity, and accuracy.[22,43] Test results are available within 30 min of sample collection.[38] Thus, the physician benefits from the test results during the initial evaluation of the patient. Specimens are not transported nor are the patients inconvenienced by referrals to clinics or laboratories. There would be no need for Gram-staining reagents, costly microscopes, and trained microscopists to interpret stained smears. The test can be performed by those who may be unskilled in laboratory procedures, and the presence or absence of gel formation is objective, absolute criteria for positive or negative test results. The LAL test correctly predicts culture results for *N. gonorrhoeae* in over 97% of the cases. Therefore, appropriate therapy can be initiated immediately and a sound, informed disposition of sexual contacts can be contemplated.

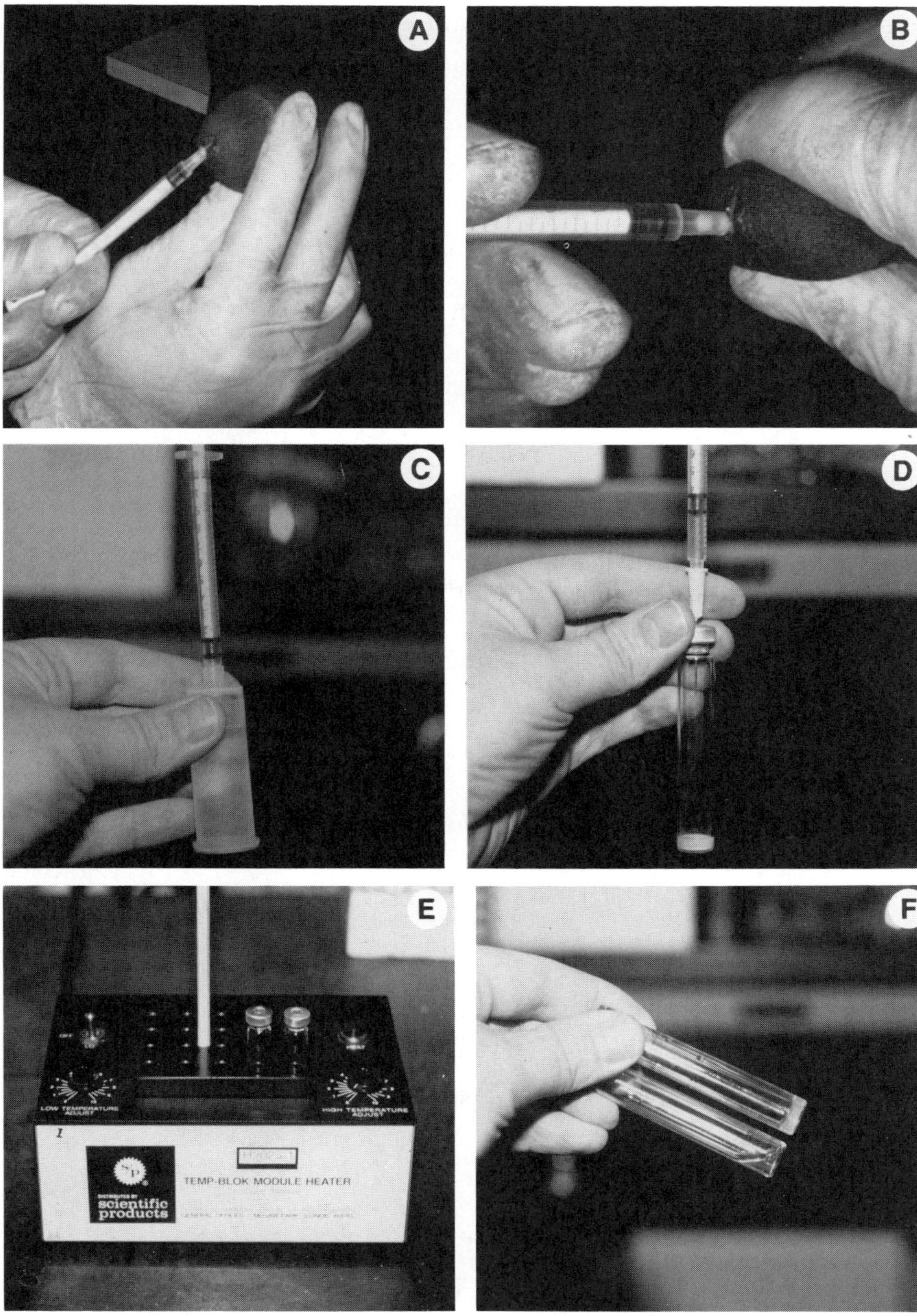

FIGURE 2. Basic procedure for using the LAL test to evaluate exudative urethritis. (A) The sample is collected at the urethral meatus using a syringe (without a needle) and gentle aspiration; (B) close-up showing about a half drop collected in the tip of the syringe (about 25 µl); (C) collection syringe with the sample affixed to dilution reservoir, creating a sealed dilution system; (D) diluted sample (0.25 ml) is transferred via the collection syringe (with needle attached) to a single-test LAL vial; (E) LAL vials incubated at 37°C for 30 min; (F) interpreting the test, positive test showing formation of an intact gel (top vial) and negative test with no gelation (bottom vial). (From Spagna, V. A. and Prior, R. B., *Am. Fam. Phys./GP*, 27(3), 227, 1983. With permission.)

TABLE 2
Comparative Results of Gram Stain and LAL Test for 1812 Male Patients with Exudative Urethritis Seen at the Columbus Health Department STD Clinic

Test parameter	Gram stain[a] (%)	LAL test (%)
Sensitivity[b]	97.1	99.3
Specificity[c]	99.3	96.8
Overall accuracy[d]	98.0	98.3

[a] Gram-stained smears read by experienced microscopists in the STD clinic.
[b] Frequency of positive test results in patients with GU.
[c] Frequency of negative test results in patients with NGU.
[d] Frequency that test will correctly predict culture results for *N. gonorrhoeae*.

TABLE 3
Study Results of the LAL Test for Evaluating Exudative Urethritis

	LAL test results (%)[a]			
No. of cases	Sensitivity	Specificity	Overall accuracy	Ref.
100	100	96	99	34
81	98	97	98	35
539	99	97	99	36
116	100	96	98	37
851	99	97	98	38
200	95	89	94	39
71	100	100	100	40
125	98	93	96	41
200	95	97	96	42

[a] Results compared to culture results.

TABLE 4
Comparative Results of Gram Stain and LAL Test (Gonoscreen) for 1100 Male Patients with Exudative Urethritis Evaluated in Five Separate Clinics in the U.S.

Test parameter	Gram stain[a] (%)	LAL test (%)
Sensitivity[b]	96.7	97.5
Specificity[c]	96.1	95.1
Overall accuracy[d]	96.5	96.9

[a,b,c,d] See Table 2 for explanation.

E. NEW DEVELOPMENTS

The major limitation of the LAL test in the evaluation of exudative urethritis was the requirement of at least 15 μl (about a half drop) of exudate to perform the test. This volume requirement was due to the type of specimen collection employed, i.e., gentle aspiration of the exudate at the urethral meatus with a needleless syringe. As a result, the utility of the LAL test was reduced, since patients with less than 15 μl of exudate were excluded.[39,48] However, testing these patients is very important, especially in the private sector where the prevalence of NGU may be high and Gram stain facilities with experienced microscopists are generally unavailable at the time of the patient's initial examination.[43]

TABLE 5
Predictive Values of Positive LAL Test for Various Prevalence Rates of GU

Theoretical prevalence (%)	Positive predictive value[a] (%)
20	83
30[b]	90
40	93
50	95
60	97
70	98

[a] Probability that a patient yielding a positive test will have GU. Computed using a sensitivity of 97.5% and a specificity of 95.1%.
[b] Relative prevalence of GU in an office practice.

Recently, the utility of the LAL test was improved by the use of pyrogen-free urogenital swabs for sample collection.[42] These dacron swabs were rendered pyrogen-free by exposure to 210°C for 30 min. A chromogenic substrate was also incorporated into the LAL test for visible color endpoint determinations following only a 10-min incubation period.[41,42] Development of a visible yellow color at the end of 10 min was scored a positive test and the lack of a visible color was scored a negative test. Use of the chromogenic substrate also eliminated the possibility of mechanically disrupting the gel during incubation. Two hundred males with various quantities of urethral exudate were evaluated, and 57 (29%) had minimal quantities of urethral exudates (less than 15 μl) and could not have been evaluated with Gonoscreen. The improved LAL test had a sensitivity and specificity of 95 and 97%, respectively, and an overall accuracy of 96% in predicting culture results. Results were not significantly different from those previously reported using the aspiration collection method. Thus, the use of urogenital swabs facilitated sample collection and increased by 29% the number of patients which could be evaluated with the LAL test. The use of a chromogenic substrate reduced the incubation time from 30 min to 10 min for test results.

V. EVALUATION OF GONOCOCCAL CERVICITIS

"Men are treated for gonorrhea; women suffer from it," according to a students' aide-mémoire.[49] Such a statement was prompted by the sharp contrast of gonorrhea in men and in women. Men usually develop symptoms such as dysuria and/or urethral discharge and seek medical attention. In contrast, early gonorrhea in women produces little or nothing by way of symptoms. Gonorrhea in women is a tale of latency, complications, chronicity, and disability.[49]

The major site of primary gonococcal infection in women is the cervix.[50] During sexual transmission, the cervix is bathed in infected male secretions. The surface pH of vaginal epithelium is too low for growth of the gonococci, except for a small area at the external os. Here the gonococci are able to attach to stratified squamous epithelium and maintain the carrier state typical of the female.[51,52] Following establishment of infection, early gonorrhea in women produces few or no symptoms. Approximately 40 to 60% of infected women will have no symptoms.[53-56] The Gram stain is less reliable in women than in men, detecting only 55 to 65% of patients who subsequently have positive cultures for *N. gonorrhoeae*.[50,56,57] A single cervical culture is also less reliable than a comparable culture in men, with reported sensitivities ranging from 80 to 93%.[50,56,58,59]

The LAL test was studied for use in the evaluation of gonococcal cervicitis. In the first report, the patient population was limited to women who had purulent endocervical secretions, and the clinical samples were collected by a cumbersome pipette/aspiration method.[60] The hypothesis was that the normal endocervix is an unfavorable environment for the commensal flora and should contain few Gram-negative bacteria. Conversely, the endocervix (squamocolumnar junction) is the site for gonococcal infection. Thus, sampling at this site should yield specimens with high concentrations of endotoxin from patients with gonococcal cervicitis. The LAL test was applied directly to clinical samples in such a way as to be highly reliable, with a sensitivity of 94% for the detection of *N. gonorrhoeae*; it could also possibly be used to assist in identifying cases of nongonococcal cervicitis.[60] Subsequent studies on cervical secretions, using aspiration methods to collect the endocervical secretions, showed that the LAL results correlated with conventional microbiological evidence of gonorrhea,[61] but that in low prevalence populations, the positive predictive value of the LAL test was not high enough for the test to function as a good screening method for the detection of gonococcal cervicitis.[62]

In a later study, the specimen collection procedure was improved by utilizing depyrogenated cotton-tipped swabs.[56] Preliminary results showed the sensitivity and specificity to be 100 and 78%, respectively. At prevalence rates of gonococcal cervicitis of 40, 20, 10, 5, and 1%, the positive predictive values of the LAL test were 97.4, 93.4, 86.4, 75.0, and 36.5%, respectively. While the predictive values are good at the higher prevalence rates, they are not sufficiently high at the lower prevalence rates to be useful clinically. However, the negative predictive values of the LAL test were 100% regardless of the prevalence of gonococcal cervicitis. Excluding an STD and assuring patients that they are not infected are at least as important as making the proper diagnosis for infected patients.[63] Therefore, the LAL test may prove beneficial in screening women in low prevalence populations to exclude the possibility of gonococcal cervicitis. There are more than 8 million screening cultures performed each year in the U.S.[64] Of these cultures, only 4.5% are positive for *N. gonorrhoeae*, and it takes at least 2 d to obtain the culture results. Failure to detect and promptly treat women with gonorrhea can lead to serious complications such as PID. In one study evaluating women seeking contraceptives, 100 women with culture-positive cervical gonorrhea were asked to report to an STD clinic for therapy. The time interval from initial screening culture to subsequent therapy was 11 d. During that time period, 7% developed definite signs and symptoms of PID and 53% had sexual intercourse.[56] The LAL test would have detected these women during their initial evaluation with a reported sensitivity of 100%,[56] and prompt treatment would have prevented the occurrence of PID and the possibility of these women unknowingly infecting their sexual partners. Thus, additional studies are warranted which would evaluate the LAL test as a rapid screening procedure for the exclusion of gonococcal cervicitis.[44]

V. EVALUATION OF SEXUAL CONTACTS AND DISSEMINATED GONORRHEA

Additional studies of the LAL test have included the testing of women exposed to men with culture-proven gonorrhea[65] and the examination of the kinetics of endotoxin in cervical secretions from women with disseminated gonorrhea.[66] In the U.S., epidemiological treatment of patients exposed to gonorrhea is recommended because of the potential complications of undetected and untreated disease.[67] However, with this approach, all individuals not infected also receive therapy, and it is estimated that they constitute 21 to 47% of those seeking treatment because of exposure to the disease.[68] Treatment of the uninfected:

1. Needlessly exposes patients to the adverse effects of antibiotics,
2. Is an unjustified expense,
3. Raises a dilemma as to the management of the sexual partners,

4. Fosters return visits by many patients for therapy to relieve their anxieties,
5. May promote selection of antibiotic-resistant bacteria,
6. Subjects many patients to the psychosocial discomfort of being treated for an STD.[63,68]

The LAL test was able to selectively exclude approximately half of culture-negative gonorrhea sexual contacts and would have spared these patients inappropriate therapy and contact tracing.[65] In the evaluation of endotoxin kinetics in women with disseminated gonorrhea, endotoxin was shown to persist in the cervical secretions for up to 20 d following appropriate antibiotic therapy. When the cervical specimens were properly diluted, the LAL test results correlated with results of conventional diagnostic methods.[66] Additional studies, however, are needed to demonstrate fully the potential utility of the LAL test in these two clinical situations.[43]

VI. SUMMARY

The LAL test has been shown to be clinically useful in the rapid evaluation of exudative urethritis in males. Unfortunately, the Gonoscreen test kit is no longer on the market due to the recent decision by Mallinckrodt to sell its *Limulus* lysate operation. However, it is hoped that another lysate producer will eventually reintroduce this test to the medical community. The newer DNA probes and ELISA test kits look promising, but additional studies are needed to determine their clinical utility. At the present time, the LAL test can compete with these new tests with regard to sensitivity, specificity, overall accuracy, simplicity, turnaround time, and cost.

Further studies are needed to determine the value of the LAL test in the female population. The ability to exclude women without gonococcal cervicitis with a negative test (negative predictive value of 100%) looks very promising. Using the LAL test as a screening test would not only exclude uninfected women but also help to identify those who are infected. This would reduce the complications associated with undetected gonorrhea and would also reduce the spread of gonorrhea by identifying the asymptomatic patient. A simple test kit designed for use in collecting endocervical sample utilizing standard collection methods (swab) has been developed[45] and is awaiting clinical evaluation.

REFERENCES

1. **Spagna, V. A. and Prior, R. B.,** *Sexually Transmitted Diseases: A Clinical Syndrome Approach*, Marcel Dekker, New York, 1985, preface.
2. **Goldsmith, M. F.,** Sexually transmitted diseaes may reverse the 'revolution', *JAMA*, 255, 1665, 1986.
3. Centers for Disease Control, Annual Summary 1984, *Morbid. Mortal. Wkly. Rep.*, 33, 57, 1986.
4. Centers for Disease Control, Summary — cases of specified notifiable diseases, United States, in *Morbid. Mortal. Wkly. Rep.*, 37, 802, 1989.
5. Centers for Disease Control, Summary of notifiable diseases, United States, *Morbid. Mortali. Wkly. Rep.*, 1988, 37, 10, 1989.
6. **Fleming, W. L., Brown, W. J., Donohue, J. F., and Barnigin, P. W.,** National survey of venereal disease treated by physicians in 1968, *JAMA*, 211, 1827, 1970.
7. **Johannisson, G., Lowhagen, G.-B., and Nilsson, S.,** *Chlamydia trachomatis* and urethritis in men, *Scand. J. Infect. Dis.*, Suppl. 32, 87, 1982.
8. **Bowie, W. R., Pollock, H. M., Forsyth, P. S., Floyd, J. F., Alexander, E. R., Wang, S. P., and Holmes, K. K.,** Bacteriology of the urethra in normal men and men with nongonococcal urethritis, *J. Clin. Microbiol.*, 6, 482, 1977.
9. **Wiesner, P. J.,** Selected aspects of the epidemiology of nongonococcal urethritis, in *Nongonococcal Urethritis and Related Infections*, Hobson, D. and Holmes, K. K., Eds., American Society for Microbiology, Washington, D.C., 1977, 9.

10. **Wilcox, R. R.,** How suitable are available pharmaceuticals for the treatment of sexually transmitted diseases? I. Conditions presenting as genital discharges, *Br. J. Vener. Dis.*, 53, 314, 1977.
11. **Volk, J. and Kraus, S. J.,** Nongonococcal urethritis — a venereal disease as prevalent as epidemic gonorrhea, *Arch. Intern. Med.*, 134, 511, 1974.
12. **Swartz, S. L.,** Diagnosis of nongonococcal urethritis, in *Nongonococcal Urethritis and Related Infections*, Hobson, D. and Holmes, K. K., Eds., American Society for Microbiology, Washington, D.C., 1977, 15.
13. **Bowie, W. R.,** Etiology and treatment of nongonococcal urethritis, *Sex. Transm. Dis.*, 5, 27, 1978.
14. **Handsfield, H. H.,** Gonorrhea and nongonococcal urethritis, recent advances, *Med. Clin. North Am.*, 62, 925, 1978.
15. **Jacobs, N. F. and Kraus, S. J.,** Gonococcal and nongonococcal urethritis in men: clinical and laboratory differentiation, *Ann. Intern. Med.*, 82, 7, 1975.
16. **Spagna, V. A.,** Urethritis syndromes, in *Sexually Transmitted Diseases: A Clinical Syndrome Approach*, Spagna, V. A. and Prior, R. B., Eds., Marcel Dekker, New York, 1985, chap. 3.
17. **Rothenberg, R. and Judson, F. N.,** The clinical diagnosis of urethral discharge, *Sex. Transm. Dis.*, 10, 24, 1983.
18. **Jordan, W. C.,** Doxycycline vs. tetracycline in the treatment of men with gonorrhea: the compliance factor, *Sex. Transm. Dis.*, 8, 105, 1981.
19. **Rothenberg, R. B., Simon, R., Chipperfield, E., and Catterall, R. D.,** Efficacy of selected diagnostic tests for sexually transmitted diseases, *JAMA*, 235, 49, 1976.
20. **Riccardi, N. B. and Felman, Y. M.,** Laboratory diagnosis in the problem of suspected gonococcal infection, *JAMA*, 242, 2703, 1979.
21. **Goodhart, M. E., Ogden, J., Zaidi, A. A., and Kraus, S. J.,** Factors affecting the performance of smear and culture tests for the detection of *Neisseria gonorrhoeae*, *Sex. Transm. Dis.*, 9, 63, 1982.
22. **Prior, R. B. and Spagna, V. A.,** Gonoscreen, *Limulus* Amebocyte Lysate Test for the Rapid Evaluation of Male Urethritis, Clinical Brochure, Mallinckrodt, St. Louis, MO, 1983.
23. **Landis, S. J., Stewart, I. O., Chernesky, M. A., Mahony, J. B., Cunningham, A. I., Grenier-Landis, M. N., and Seidelman, W. E.,** Value of the Gram-stained urethral smear in the management of men with urethritis, *Sex. Transm. Dis.*, 15, 78, 1988.
24. **Burns, M., Rossi, P. H., Cox, D. W., Edwards, T., Kramer, M., and Kraus, S. J.,** A preliminary evaluation of the Gonozyme test, *Sex. Transm. Dis.*, 10, 180, 1983.
25. **Schachter, J., McCormack, M., Smith, R. F., Parks, R. M., Bailey, R., and Ohlin, A. C.,** Enzyme immunoassay for diagnosis of gonorrhea, *J. Clin. Microbiol.*, 19, 57, 1984.
26. **Stamm, W. E., Cole, B., Fennell, C., Bonin, P., Armstrong, A. S., Herrmann, J. E., and Holmes, K. K.,** Antigen detection for the diagnosis of gonorrhea, *J. Clin. Microbiol.*, 19, 399, 1984.
27. **Granato, P. A. and Roefaro, M.,** Comparative evaluation of enzyme immunoassay and culture for the laboratory diagnosis of gonorrhea, *Am. J. Clin. Pathol.*, 83, 613, 1985.
28. **Jones, M. F., Smith, T. F., Houglum, A. J., and Herrmann, J. E.,** Detection of *Chlamydia trachomatis* in genital specimens by the Chlamydiazyme test, *J. Clin. Microbiol.*, 20, 465, 1984.
29. **Totten, P. A., Holmes, K. K., Handsfield, H. H., Knapp, J. S., Perrine, P. L., and Falkow, S.,** DNA hybridization technique for the detection of *Neisseria gonorrhoeae* in men with urethritis, *J. Infect. Dis.*, 148, 462, 1983.
30. **Granato, P. A. and Franz, M. R.,** Evaluation of a prototype DNA probe test for the noncultural diagnosis of gonorrhea, *J. Clin. Microbiol.*, 27, 632, 1989.
31. **Blair, D. J., Gonzalez, D., Kranig-Brown, D., Fletcher, D. E., Kohne, D. E., and Shaw, S. B.,** Evaluation of a simplified nonisotopic DNA probe assay for *Chlamydia trachomatis*, *Abstracts, 29th Interscience Conference on Antimicrobial Agents and Chemotherapy*, American Society for Microbiology, Washington, D. C., 1989.
32. **Gaydos, C. A., Holland, S. M., and Quinn, T. C.,** Detection of *Chlamydia trachomatis*, *Chlamydia psittaci*, and *Chlamydia pneumoniae* TWAR by polymerase chain reaction, *Abstracts, 29th Interscience Conference on Antimicrobial Agents and Chemotherapy*, American Society for Microbiology, Washington, D.C., 1989.
33. **Granato, P. A. and Murfitt, K.,** Evaluation of the pathfinder EIA test for detecting chlamydia in endocervical specimens, *Abstracts, 29th Interscience Conference on Antimicrobial Agents and Chemotherapy*, American Society for Microbiology, Washington D. C., 1989.
34. **Spagna, V. A., Prior, R. B., and Perkins, R. L.,** Rapid presumptive diagnosis of gonococcal urethritis in men by the limulus lysate test, *Br. J. Vener. Dis.*, 55, 179, 1979.
35. **Prior, R. B. and Spagna, V. A.,** Comparative evaluation of the tube and microdilution *Limulus* lysate techniques for rapid presumptive diagnosis of gonococcal urethritis in men, *J. Clin. Microbiol.*, 11, 340, 1980.
36. **Spagna, V. A. and Prior, R. B.,** The limulus amebocyte lysate assay, *Am. Fam. Phys./GP*, 22, 125, 1980.
37. **Prior, R. B. and Spagna, V. A.,** Response of several *Limulus* amoebocyte lysates to native endotoxin in gonococcal and nongonococcal urethral exudates from human males, *J. Clin. Microbiol.*, 13, 167, 1981.
38. **Prior, R. B. and Spagna, V. A.,** Application of a *Limulus* test device in rapid evaluation of gonococcal and nongonococcal urethritis in males, *J. Clin. Microbiol.*, 14, 256, 1981.

39. **Chapel, T. A., Adcock, M., Smith, B., Barnes, T., and Goodman, M.,** Evaluation of the *Limulus* amebocyte lysate assay for presumptive diagnosis of gonorrhea in men at a clinic for sexually transmitted diseases, *Sex. Transm. Dis.*, 8, 175, 1981.
40. **Nachum, R. and Christiansen, R. O.,** Rapid presumptive diagnosis of gonococcal urethritis in males: evaluation of a prototype limulus lysate test kit, *Med. Lab. Sci.*, 39, 83, 1982.
41. **Prior, R. B. and Spagna, V. A.,** Rapid evaluation of gonococcal and nongonococcal urethritis in men with *Limulus* amoebocyte lysate and a chromogenic substrate, *J. Clin. Microbiol.*, 17, 485, 1983.
42. **Prior, R. B. and Spagna, V. A.,** Improved utility of Gonoscreen, a *Limulus* amoebocyte lysate assay, in the evaluation of urethral discharges in men, *J. Clin. Microbiol.*, 22, 141, 1985.
43. **Prior, R. B.,** Bacterial endotoxin detection by the *Limulus* amoebocyte lysate test, in *Antigen Detection to Diagnose Bacterial Infections*, Vol. 2, Kohler, R. E., Ed., CRC Press, Boca Raton, FL, 1986, 250.
44. **Spagna, V. A. and Prior, R. B.,** Limulus lysate assay for gonorrhea, *Am. Fam. Phys./GP*, 27, 227, 1983.
45. **Prior, R. B. and Spagna, V. A.,** U.S. Patent 4,376,634, 1983.
46. **Vecchio, T. J.,** Predictive value of a single diagnostic test in unselected populations, *N. Engl. J. Med.*, 274, 1171, 1966.
47. **Kleris, G. S. and Arnold, A. J.,** Differential diagnosis of urethritis: predictive value and therapeutic implications of the urethral smear, *Sex. Transm. Dis.*, 8, 110, 1981.
48. **Judson, F. N., Werness, B. A., and Shahan, M. R.,** Lack of utility of a *Limulus* amoebocyte lysate assay in the diagnosis of urethral discharges in men, *J. Clin. Microbiol.*, 21, 152, 1985.
49. **Morton, R. S.,** *Gonorrhoea*, W. B. Saunders, London, 1977, 94.
50. **Dans, P. E.,** Gonococcal anogenital infection, *Clin. Obstet. Gynecol.*, 18, 103, 1975.
51. **Evans, B. A.,** Ultrastructural study of cervical gonorrhea, *J. Infect. Dis.*, 136, 248, 1977.
52. **Prior, R. B. and Spagna, V. A.,** Cervicitis syndromes, in *Sexually Transmitted Diseases: A Clinical Syndrome Approach*, Spagna, V. A. and Prior, R. B., Eds., Marcel Dekker, New York, 1985, chap. 4.
53. **Wiesmer, P. J.,** Gonorrhea, *Cutis*, 27, 249, 1981.
54. **Jha, P. K., Singh, G., Kaur, P., and Sharma, D.,** Unsuspected gonococcal infection in female patients, *Br. J. Vener. Dis.*, 54, 324, 1978.
55. **Little, R. G. and Hartsfield, M.,** Incidence of asymptomatic gonorrhea in women, *J. Fam. Pract.*, 7, 597, 1978.
56. **Spagna, V. A., Prior, R. B., and Sawaya, G. A.,** Sensitivity, specificity, and predictive values of the *Limulus* lysate assay for detection or exclusion of gonococcal cervicitis, *J. Clin. Microbiol.*, 16, 77, 1982.
57. **Young, H.,** Advances in routine laboratory procedures for the diagnosis of gonorrhoea, in *Recent Advances in Sexually Transmitted Diseases*, Harris, Ed., Churchill Livingstone, Edinburgh, 1981, 59.
58. **Caldwell, J. G., Price, E. V., Pazin, G. J., and Cornelius, C. E.,** Sensitivity and reproducibility of Thayer-Martin culture medium in diagnosing gonorrhea in women, *Am. J. Obstet. Gynecol.*, 109, 463, 1971.
59. **Holmes, K. K.,** The Chlamydia epidemic, *JAMA*, 245, 1718, 1981.
60. **Spagna, V. A., Prior, R. B., and Perkins, R. L.,** Rapid presumptive diagnosis of gonococcal cervicitis by the Limulus lysate assay, *Am. J. Obstet. Gynecol.*, 137, 595, 1980.
61. **Young, H., Sarafian, S. K., and McMillan, A.,** Reactivity of the Limulus lysate assay with uterine cervical secretions — a preliminary evaluation, *Br. J. Vener. Dis.*, 57, 200, 1981.
62. **Hainer, B. L., Danylchuke, P., Cooper, J., and Weart, C. W.,** Limulus lysate assay in detection of gonorrhea in women from a low-incidence population, *Am. J. Obstet. Gynecol.*, 144, 67, 1982.
63. **Kraus, S. J.,** Semiquantitaion of urethral polymorphonuclear leukocytes as objective evidence of nongonococcal urethritis, *Sex. Transm. Dis.*, 9, 52, 1982.
64. Sexually Transmitted Disease (STD) Statistical Letter, Issue no. 129, Centers for Disease Control, Atlanta, 1979, 9.
65. **Prior, R. B. and Spagna, V. A.,** Rapid evaluation of female patients exposed to gonorrhea by use of the *Limulus* lysate test, *J. Clin. Microbiol.*, 16, 487, 1982.
66. **Spagna, V. A. and Prior, R. B.,** Application of the Limulus lysate assay in the evaluation of disseminated gonorrhea in women, *Sex. Transm. Dis.*, 8, 18, 1981.
67. Centers for Disease Control, 1989 STD Treatment Guidelines, *Morbid. Mortal. Wkly. Rep.*, 38, 23, 1989.
68. **Johnson, R. E.,** Epidemiologic and prophylactic treatment of gonorrhea: a decision analysis review, *Sex. Transm. Dis.*, 6, 159, 1979.

Chapter 9

EVALUATION OF PYOGENIC ARTHRITIS

Ronald J. Elin

TABLE OF CONTENTS

I. Introduction 112

II. Clinical Utility of the LAL Test for Diagnosis of Pyogenic Arthritis 112

III. Discussion 113

References 114

I. INTRODUCTION

The clinical and laboratory diagnosis of pyogenic arthritis is frequently difficult. The clinical features of pyogenic arthritis are not specific for the disease and may be mistaken for nonseptic disease processes.[1] The physician frequently relies on the laboratory analysis of aspirated synovial fluid for the diagnosis of pyogenic arthritis. However, a Gram stain of synovial fluid from a patient with pyogenic arthritis may be positive for the organism only in 50% of the cases[2] and culture results may be negative.[2-4] The white blood cell (WBC) count and differential of cells in the synovial fluid are usually not specific for a pyogenic process because similar results may occur in synovial fluid from patients with nonseptic inflammatory joint disease.[5]

Four studies have evaluated the clinical utility of the *Limulus* amoebocyte lysate (LAL) test for the detection of bacterial endotoxin in pyogenic arthritis and nonseptic inflammatory joint disease.[6-9] This chapter is an attempt to summarize these studies and assess the current clinical value of the LAL test for the diagnosis of pyogenic arthritis.

II. CLINICAL UTILITY OF THE LAL TEST FOR THE DIAGNOSIS OF PYOGENIC ARTHRITIS

Four studies have compared the diagnosis of a joint disorder and the LAL test result with synovial fluid analysis.[6-9] In three of these studies the synovial fluid specimens were cultured;[6-8] the fourth study was limited to nonseptic inflammatory joint disease without culture results.[9] A comparison of LAL and culture results from synovial fluid among the three studies is presented in Table 1.

Results for six synovial fluid specimens were reported by Tuazon et al.[6] The LAL test detected 1 ng of purified endotoxin per milliliter. Chloroform extraction was used on synovial fluids with a high protein content. Four synovial fluids grew Gram-negative organisms and gave a positive LAL test result. Two synovial fluids grew Gram-positive organisms, but only one was positive with the LAL test (Table 1). The authors suggested that the LAL test had merit for the early detection of Gram-negative septic arthritis.

Elin et al.[7] studied 71 specimens of synovial fluid from 46 patients with a spectrum of joint diseases. The LAL test was similar to that used in the study by Tuazon et al. and could detect 1 ng of a purified endotoxin per milliliter. In this study, synovial fluid was collected with and without heparin and the LAL test was performed on synovial fluid with and without chloroform extraction. Significantly more positive tests for endotoxin occurred when the synovial fluid was collected with heparin and extracted with chloroform prior to LAL testing. The results in Table 1 represent the first synovial fluid specimen from each of the 46 patients. Organisms grew in three synovial fluids (one Gram-negative and two Gram-positive organisms), and all gave a positive LAL test result. Of 43 synovial fluid specimens with no growth, 18 (41.9%) were positive for the LAL test. In this study, a positive LAL test result correlated significantly with an elevated WBC count ($p < 0.0005$), an elevated absolute neutrophil count ($p < 0.0005$), and a decreased concentration of glucose ($p < 0.005$) in synovial fluid. These authors concluded that a positive LAL test result with synovial fluid is nonspecific for a septic process, but a negative LAL test result indicated nonseptic joint disease.

Cesario et al.[8] studied synovial fluid specimens from 69 patients. The *Limulus* lysate for the study was purchased from Mallinckrodt, Inc. and was sensitive to 0.0125 ng of a purified endotoxin per milliliter. If the synovial fluid were positive with the LAL test, serial dilutions were made with pyrogen-free water to quantify the endotoxin. The authors reported that 1 synovial fluid specimen contained 128 ng of endotoxin per milliliter. The next highest concentration was 8 ng of endotoxin per milliliter. Four of six synovial fluid specimens that grew Gram-negative organisms gave a positive LAL test result (Table 1). One of two synovial fluid specimens that grew Gram-positive organisms gave a positive LAL test result. Of the 61 sterile

TABLE 1
Comparison of LAL and Culture Results for Synovial Fluid

	LAL result	Culture results: Gram neg	Gram pos	No growth	Total	Ref.
	Pos	4	1	0	5	6
	Neg	0	1	0	1	
	Pos	1	2	18	21	7
	Neg	0	0	25	25	
	Pos	4	1	3	8	8
	Neg	2	1	58	61	
Total	Pos	9	4	21	34	
	Neg	2	2	83	87	

synovial fluid specimens, only 3 gave a positive LAL test result. These authors concluded that the LAL test appeared to have specificity but lacked sensitivity.

Wagener et al. studied synovial fluid from 100 patients with inflammatory rheumatic diseases and 30 controls. These authors used a quantitative LAL assay. The mean concentration ± SD for the synovial fluid from the 30 control subjects was 3.6 ± 1.9 pg of endotoxin per milliliter. The mean endotoxin concentration in synovial fluid from patients with a diagnosis of rheumatoid arthritis, ankylosing spondylitis, or Crohn's disease was significantly greater ($p < 0.05$) than the control group. Thus, the apparent concentration of endotoxin increased in the synovial fluid of patients with noninfectious inflammatory rheumatic diseases.

III. DISCUSSION

Results vary among the four studies in which the LAL test was applied to specimens of synovial fluid and do not give a clear indication of clinical utility for the diagnosis of septic joint disease. As indicated, the culture of synovial fluid may be negative with septic arthritis.[2-4] Thus, the culture result could not be considered the "gold standard" for comparison of LAL test results with a septic process. Nonetheless, LAL test results have commonly been compared with culture results. However, the error is primarily limited to a false negative culture result. Among the 3 studies, 11 specimens of synovial fluid grew Gram-negative organisms (presumably contained endotoxin), and 9 of the specimens gave positive LAL test results. Thus, the sensitivity of the LAL test for the diagnosis of septic arthritis due to Gram-negative organisms is 82%. It is of interest that 4 of 6 (67%) synovial fluid specimens that grew Gram-positive organisms (presumably did not contain endotoxin) gave positive LAL test results. Among the 3 studies 104 synovial fluid specimens were sterile, but 21 gave a positive LAL test result. Thus, the specificity of the LAL test for sterile synovial fluid is 80% (Table 1). Presuming some false-negative culture results, the "true" specificity should be higher.

Can synovial fluid from noninfectious inflammatory joint disease give a positive LAL test result? This is the key question for establishing the specificity of the test. Nucleic acids can affect a positive LAL test result.[10] Noninfectious inflammatory synovial fluid specimens may have a markedly elevated WBC count with an increase in the nucleic acid concentration. This hypothesis is supported by two of the studies.[7,9] The study by Elin et al. found that the LAL test was frequently positive with synovial fluid from patients with a nonseptic disease processes (5 of 11 with gout, 5 of 6 with rheumatoid arthritis, and 2 of 4 with Reiter's syndrome). Five synovial fluid specimens from patients with osteoarthritis were negative with the LAL test, and all had markedly lower WBC counts compared to other joint diseases. The study by Wagener et al.[9] supports these findings since significantly greater concentrations of endotoxin were detected in the synovial fluid from patients with nonseptic inflammatory rheumatic diseases compared with controls. Thus, evidence suggests that the specificity of the test may be impaired by false-positive results due to compounds other than endotoxin.

The variability among these studies may be due to differences in the *Limulus* lysate preparation, specimen collection, and treatment of the synovial fluid prior to testing. The lysate used by Cesario et al. was 80-fold more sensitive for the detection of endotoxin than the lysate used in the studies by Tuazon et al. and Elin et al. Contrary to this difference in LAL sensitivity, Elin et al. found that 42% of sterile synovial fluid specimens gave a positive LAL result, compared to only 5% found by Cesario et al. Thus, LAL sensitivity for detection of endotoxin in these two studies conflicts with the anticipated result with sterile synovial fluid. These results could be explained by a greater percentage of false-negative culture results or by the LAL reacting with compounds other than endotoxin in the study by Elin et al. In addition, 1 of the synovial fluid specimens (culture negative or results unknown) was reported by Cesario et al. to contain 128 ng of endotoxin per milliliter, a high concentration. In this author's opinion, this high result suggests that a compound other than endotoxin reacted with the lysate.

The study by Elin et al. found that the synovial fluid specimen should be collected with heparin for anticoagulation and extracted with chloroform prior to testing to maximize the probability of a positive LAL test result. In three of the studies they did not anticoagulate the specimen with heparin,[6,8,9] and in two they did not treat the joint fluid with chloroform or other methods to "liberate" endotoxin.[8,9] A study that optimizes specimen collection and treatment of serum prior to testing with state-of-the-art LAL technology is needed to define the clinical value of the test.

REFERENCES

1. **Kellgren, J. H., Ball, J., Fairbrother, R. W., and Barnes, K. L.,** Suppurative arthritis complicating rheumatoid arthritis, *Br. Med. J.*, 1, 1193, 1958.
2. **Goldenburg, D. L., Brandt, K. D., Cohen, A. S., and Cathcart, E. S.,** Treatment of septic arthritis. A comparison of needle aspiration and surgery as initial modes of treatment, *Am. J. Med.*, 60, 369, 1976.
3. **Russeli, A. S. and Ansell, B. M.,** Septic arthritis, *Ann. Rheum. Dis.*, 31, 40, 1972.
4. **Kelly, P. J.,** Bacterial arthritis in the adult, *Orthop. Clin. North Am.*, 6, 973, 1975.
5. **Ward, J. R. and Atcheson, S. G.,** Infectious arthritis, *Med. Clin. North Am.*, 61, 313, 1977.
6. **Tuazon, C. U., Perez, A. A., Elin, R. J., and Sheagren, J. N.,** Detection of endotoxin in cerebrospinal and joint fluids by limulus assay, *Arch. Intern. Med.*, 137, 55, 1977.
7. **Elin, R. J., Knowles, R., Barth, W. F., and Wolff, S. M.,** Lack of specificity of the limulus lysate test in the diagnosis of pyogenic arthritis, *J. Infect. Dis.*, 137, 507, 1978.
8. **Cesario, T., Jason, M., Andrews, B. S., Shanbron, E., Juranas, D., Katz, L., and Friou, G. J.,** Limulus assay for bacterial endotoxin in synovial fluid, *Ann. Rheum. Dis.*, 42, 571, 1983.
9. **Wagener, P., Busch, M., and Brunkhorst, R.,** Endotoxin determination of inflammatory rheumatic diseases, *Clin. Rheumatol.*, 6(Abstr.), 110, 1987.
10. **Elin, R. J. and Utter, A.,** Positive limulus amebocyte lysate reactions with polyriboinosinic acid-polyribocytidylic acid, *J. Clin. Microbiol.*, 12, 502, 1980.

Chapter 10

EVALUATION OF OTITIS MEDIA

Thomas F. DeMaria and David J. Lim

TABLE OF CONTENTS

I. Introduction 116
  A. Acute Otitis Media 116
  B. Otitis Media with Effusion (OME) 117
  C. Evaluation of Otitis Media 117

II. *Limulus* Amoebocyte Lysate Test 117

References 119

I. INTRODUCTION

Otitis media is predominantly a disease of infants and young children and a major worldwide health problem. Otitis media is defined as an inflammation of the middle ear and can be classified into a spectrum of different clinical types of diseases. The two types most pertinent to this chapter are acute otitis media and otitis media with effusion (OME). The etiology, epidemiology, and relatedness of these clinical types of otitis media continue to be subjects of intense investigation.

Recent studies indicate that by 2 to 3 years of age, 61 to 71% of all children have had an episode of otitis media. The impact of this disease on medical practice is immense. A study in Boston indicated that, for children during the first 3 years of life, approximately one physician visit in three and three fourths of follow-up visits were for middle ear disease.[1]

Recent advances in the epidemiology of this condition clearly indicate that the age at the first episode is an important risk-predictor of recurring disease. Children who experience two or more episodes of otitis media during the first year of life will have twice as many subsequent episodes as the child who has had one or no episodes during this time period.

Most of the serious complications of otitis (e.g., mastoiditis, meningitis) have disappeared with the use of antibiotics. However, the relationship of the mild conductive hearing loss accompanying otitis media to language and learning development is only just beginning to be explored. Recent developments in this area suggest a relationship between an otitis media-induced hearing loss and parent-child bonding, vocabulary deficits, language learning deficits, aberrant behavioral patterns, and reduced educational achievement.[2-4]

A description of the clinical syndromes, their etiologies, epidemiology, diagnosis, and treatment is outlined in this chapter.

A. ACUTE OTITIS MEDIA

The hallmark of acute otitis media is the rapid onset of signs and symptoms of middle ear inflammation, including fever, pain, and recent onset of irritability in the small child. The tympanic membrane (ear drum) is usually full or bulging and erythematous. Fluid (middle ear effusion) may be present in the middle ear. Synonyms for this condition are suppurative or purulent otitis media. Following an episode of acute otitis media, a middle ear effusion can persist for several months.

Most cases of acute otitis media have been associated with the recovery of pathogenic bacteria from middle ear exudates. Numerous studies during the past 40 years have documented that two organisms, *Streptococcus pneumoniae* and *Hemophilus influenzae*, are the principal etiological agents.[5] Among the types of pneumococci causing otitis media, 8 serotypes account for approximately 80% of all cases.[6] Serologically nontypable strains of *H. influenzae* account for over 90% of all *H. influenzae* otitis isolates.[5] This is in contrast to all other pediatric infections where type b isolates are the predominate strains. *Moraxella (Branhamella) catarrhalis* has also become a significant otitis media pathogen during the past 10 years.[7]

Nonsurgical management of acute otitis media includes antimicrobials, antihistamines, decongestants, mucolytic agents, and anti-inflammatory agents. Definitive studies regarding the efficacy of these treatments are not available. The predilection of certain strains of bacteria which induce otitis media to acquire resistance to various antibiotics requires a constant regular evaluation of the susceptibility of each pathogenic species to a particular antibiotic. This represents a major effort since most cases of otitis media are treated empirically, without aspiration and culture of the middle ear fluids for bacteria. There are no standardized data addressing the relationship of age, race, sex, anatomic, or environmental factors and their effect on antimicrobial and other forms of medical treatment.

Surgical management of acute otitis media includes myringotomy (surgical incision of the tympanic membrane) and aspiration of middle ear effusion alone or together with tympanostomy tube insertion. These combined treatments are recommended for acute otitis media with suppurative complications, severe otalgia requiring immediate relief, and microbiologic diagnosis.

B. OTITIS MEDIA WITH EFFUSION (OME)

A middle ear effusion that persists for longer than 3 months following an episode of acute otitis media is called chronic OME.[1] It has been reported that as many as 97% of the cases of OME had histories of prior acute otitis media.[8] The presence of a relatively asymptomatic middle ear effusion has many synonyms, e.g., "secretory", "nonsuppurative", or "serous otitis media", but the most acceptable term is OME. The most important distinction between this type of disease and acute otitis media is that the signs and symptoms of acute infection (otalgia, fever, etc.) are not present in OME. However, hearing loss may be present in both conditions.

The pathogenesis of OME remains controversial, and several major hypotheses have been proposed. Susceptibility to OME parallels the anatomical maturation of the eustachian tube muscles and cartilage.[9] Patients with increased eustachian tube compliance (lack of stiffness) are prone to developing OME.[10] Poor tubal function is thought to result in an increased negative pressure in the middle ear, which promotes the transudation of fluid into the middle ear. Negative pressure may also increase the chance for an infection of the middle ear, by promoting the aspiration of a bolus of mucus containing bacteria from the nasopharynx into the middle ear.

Although the middle ear effusions from cases of OME were originally presumed to be sterile, recent data have indicated that 22 to 40% of the middle ear effusions show culturable bacteria, and 80% of the samples demonstrate bacteria on Gram stain.[11] The predominate bacteria are the same as the strains isolated from cases of acute otitis media. The exact role of bacteria in the pathogenesis of OME has not been elucidated.

Additional factors that have been proposed as contributing to the pathogenesis of OME include upper respiratory tract virus infections, allergy, hypersecretion by the middle ear epithelium, and inadequate immune protection. How these various factors, individually or acting in concert, contribute to the pathogenesis of OME is still largely unknown.

As with acute otitis, nonsurgical management of OME includes antimicrobials, antihistamines, decongestants, mucolytic agents, and anti-inflammatory agents. Failure of these conservative treatment methods to eradicate the disease usually results in surgical intervention. Myringotomy and the insertion of tympanostomy tubes to ventilate the middle ear space is the surgical procedure requiring general anesthesia that is most commonly performed in children. Over 1 million children receive tympanostomy tubes annually.

C. EVALUATION OF OTITIS MEDIA

The evaluation of otitis media is based primarily on inspection of the tympanic membrane while varying the air pressure in the auditory canal by means of a pneumotoscope. Pneumotoscopy is the minimum diagnostically acceptable procedure. An assessment of middle ear pressure and compliance by tympanometry and measurement of hearing loss by means of pure tone audiometry are important additional diagnostic tools.

Unlike the other clinical situations described in this volume, laboratory evaluation of patient specimens is not required or recommended for diagnosis or treatment. Antibiotics are prescribed empirically based on clinical epidemiological data. Middle ear effusions are occasionally collected by scientists for research purposes, usually at the time of surgery for tympanostomy tube insertion.

II. *LIMULUS* AMOEBOCYTE LYSATE TEST

The fact that a majority of the culture-negative middle ear effusions during cases of otitis media demonstrate bacteria on Gram stain has prompted investigations into the role of biologically active microbial cell wall components in the pathogenesis of otitis media with effusion. It has been postulated that microbial cell wall components that are not eradicated by local host defense mechanisms may persist in the middle ear following resolution of the acute infection phase through effective microbial treatment.[12] One such substance is endotoxin, the lipopolysaccharide (LPS) cell wall component of Gram-negative bacteria such as the otitis media pathogens *H. influenzae* and *M. catarrhalis*.

Numerous investigators have detected endotoxin in the middle ear effusions (MEEs) from patients with otitis media by means of the *Limulus* amoebocyte lysate (LAL) test. Bernstein et al.[13] detected endotoxin in the MEEs, cholesteatomas, and granulation tissue of 13 specimens from 31 patients with chronic otitis media. Only 7% of the sterile MEEs in this study contained endotoxin. DeMaria et al.[14] found endotoxin in 71 of 89 (80%) middle ear fluids from patients with chronic otitis media. Endotoxin was present in 67% of the culture-negative MEEs in this study. The mean concentration was 12.9 ng/ml.

The discrepancy between these two studies cannot be fully explained, but may be due to the fact that two different patient populations were being studied and different commercial sources of lysate were used, which resulted in different sensitivities. The *Limulus* lysate used in the study by DeMaria et al. had previously been shown to be more sensitive to native endotoxin than lysate prepared by using chloroform extraction.[15] Subsequent studies by Iino et al.[16] and Nonomura et al.[17] clearly indicate that the incidence of endotoxin in sterile MEEs from patients with otitis media is greater than 65%.

A recent report by Palva and Lehtinen[18] suggested that the endotoxin positivity demonstrated in the MEEs is a nonspecific reaction due to the presence of aggregated immunoglobulins (IgG and IgA) in the MEEs. We have attempted to repeat their study using reagents obtained from the same source (Sigma). Our data[19] indicate that the commercially available IgG and IgA used in their study were probably grossly contaminated with pyrogens. No mention is made in their report of boiling their preparations prior to assay, which prohibits establishing with certainty what substance is responsible for the positive gelation observed in their study.

Endotoxin is a biologically active material and a potent inducer of inflammation, as well as a modulator of the immune response. In addition, endotoxin is capable of interaction with complement and components of the clotting system. Endotoxin has been shown to effect the release of various vasoactive amines and other mediators of inflammation by direct interaction with macrophages, polymorphonuclear leukocytes, and other cell types.[20] The mechanisms by which endotoxin exerts its biologic activities remain unknown. There is a consensus that individual biologic activities are caused by endogenous mediators whose formation and/or release follow the interaction of endotoxin with the appropriate targets. Macrophages have been suggested by Freudenberg et al.[21] as central effector cells in endotoxin-mediated host responses.

In view of the preceeding discussion, it is not surprising that experimental studies conducted so far indicate that endotoxin induces severe inflammatory changes in the middle ear and eustachian tube.[12,17,22,23]

Finally, the LAL test has been crucial to one new avenue of research, namely the assessment of the ability of endotoxin to pass from the middle to the inner ear, presumably via the round window membrane. A high incidence of sensorineural hearing loss and separation of the otoconial membrane have been reported following episodes of otitis media.[24] Whether bacterial antigens or toxins can penetrate the round window membrane and damage the inner ear has not been determined. In a preliminary report, Spandow et al.[25] showed that the instillation of endotoxin in the middle ear caused effusion formation and impairment of inner ear function. Auditory brain stem response testing indicated a hearing loss in the 6- to 31.5-kHz range that became maximal 3 to 4 d after injection of endotoxin into the middle ear. This study indicates damage to the basal portion of the cochlea near the round window. No direct measurement of the endotoxin in the inner ear was undertaken as part of this study.

Recently, however, Kawauchi et al.[26] in an experimental study demonstrated that endotoxin is demonstrable in the perilymph in chinchillas 12 h after injection of the middle ear and persists for at least 14 d. Endotoxin in the perilymph was measured by means of a chromogenic LAL test. The results from this study indicate that endotoxin present in the middle ear can permeate the inner ear and induce a potent inflammatory response leading to sensory cell degeneration and possibly other sequelae associated with otitis media.

Although it is not used for patient diagnosis or clinical evaluation, the LAL test continues to be an important research tool for clinical and experimental studies of chronic otitis media and its sequelae. It will also play an important role in future research studies of this clinical syndrome.

REFERENCES

1. **Teele, D. W., Rosner, B. A., and Klein, J. O.,** Epidemiology of otitis media in children, *Ann. Otol. Rhinol. Laryngol.*, 89(Suppl. 68), 5, 1980.
2. **Teele, D. W., Klein, J. O., and Rosner, B. A.,** Otitis media with effusion during the first three years of life and development of speech and language, *Pediatrics*, 74, 282, 1984.
3. **Robert, J. E., Sanyal, M. A., and Burchinal, M. R.,** Otitis media in early childhood and its relationship to later verbal and academic performance, *Pediatrics*, 78, 423, 1986.
4. **Lous, J., Fiellau-Nikolajsen, M., Jeppesen, A. L.,** Secretory otitis media and verbal intelligence: a six year prospective case-control study, in *Recent Advances in Otitis Media*, Lim, D. J., Bluestone, C. D., Klein, J. O., and Nelson, J. D., Eds., B.C. Decker, Toronto, 1988, 393.
5. **Giebink, G. S.,** The microbiology of otitis media, *Pediatr. Infect. Dis. J.*, 8, S18, 1989.
6. **Austrian, R.,** Epidemiology of pneumococcal capsular types causing pediatric infections, *Pediatr. Infect. Dis. J.*, 8, S21, 1989.
7. **Shurin, P. A., Marchant, C. D., Kim, C. H., Van Hare, G. F., Johnson, C. E., Tutihasi, M. A., and Knapp, L. J.,** Emergence of beta-lactamase-producing strains of *Branhamella catarrhalis* as important agents of acute otitis media, *Pediatr. Infect. Dis. J.*, 2, 34, 1983.
8. **Giebink, G. S., Mills, E. J., and Huff, J. S.,** The microbiology of serous and mucoid otitis media, *Pediatrics*, 63, 915, 1979.
9. **Casselbrant, M. L., Okeowo, P. A., and Flaherty, M. R.,** Prevalence and incidence of otitis media in a group of preschool children in the United States, in *Recent Advances in Otitis Media*, Lim, D. J., Bluestone, C. D., Klein, J. O., and Nelson, J. D., Eds., B.C. Decker, Toronto, 1988, 393.
10. **Bluestone, C. D. and Berry, Q. C.,** Concepts on the pathogenesis of middle ear effusions, *Ann. Otol. Rhinol. Laryngol.*, 85(Suppl. 25), 182, 1976.
11. **Liu, Y. S., Lim, D. J., Lang, R. W., and Birck, H. G.,** Chronic middle ear effusions: immunochemical and bacteriological investigation, *Arch. Otolaryngol.* 101, 278, 1975.
12. **DeMaria, T. F., Briggs, B. R., Lim, D. J., and Okazaki, N.,** Experimental otitis media with effusion following middle ear inoculation of nonviable *Hemophilus influenzae*, *Ann. Otol. Rhinol. Laryngol.*, 93, 52, 1984.
13. **Bernstein, J. M., Praino, M. D., and Neter, E.,** Detection of endotoxin in ear specimens from patients with chronic otitis media by means of limulus amebocyte test, *Can. J. Microbiol.*, 26, 645, 1980.
14. **DeMaria, T. F., Prior, R. B., Briggs, B. R., Lim, D. J., and Birck, H. G.,** Endotoxin in middle ear effusion from patients with chronic otitis media with effusion, *J. Clin. Microbiol.*, 20, 15, 1984.
15. **Prior, R. B. and Spagna, V. A.,** Response of several *Limulus* amoebocyte lysates to native endotoxin present in gonococcal and nongonococcal urethial exudates from human males, *J. Clin. Microbiol.*, 13, 167, 1981.
16. **Iino, Y., Kaneko Y., and Takasaka, J.,** Endotoxin in middle ear effusions tested with *Limulus* assay, *Acta Otolaryngol. (Stockholm)*, 100, 42, 1985.
17. **Nonomura, N., Nakano, Y., Satoh, Y., Fujioka, O., Niijima, H., and Fujita, M.,** Otitis media with effusion following inoculation of *Haemophilus influenzae* type b endotoxin, *Arch. Otorhinolaryngol.*, 243, 31, 1986.
18. **Palva, T. and Lehtinen, T.,** Pneumococcal antigens and endotoxin in effusions from patients with secretory otitis media, *Int. J. Pediatr. Otorhinolaryngol.*, 14, 123, 1987.
19. **DeMaria, T. F. and Prior, R. B.,** unpublished data, 1989.
20. **Morrison, D. C. and Ulevitch, R. J.,** The effects of bacterial endotoxin on host mediation systems, *Am. J. Pathol.*, 93, 527, 1978.
21. **Freudenberg, M. A., Keppler, D., and Galanos, C.,** Requirement for lipopolysaccharide-responsive macrophages in galactosamine-induced sensitization to endotoxin, *Infect. Immun.*, 51, 891, 1986.

22. **DeMaria, T. F., Yamaguchi, T., and Lim, D. J.,** Quantitative cytological and histological changes in the middle ear after the injection of nontypable *Hemophilus influenzae* endotoxin, in *Recent Advances in Otitis Media*, Lim, D. J., Bluestone, C. D., Klein, J. O., and Nelson, J. D., Eds., B.C. Decker, Toronto, 1988, 320.
23. **Ohashi, Y., Nakai, Y., Ikeoka, H., Koshimo, H., and Esaki, Y.,** Effects of bacterial endotoxin on the ciliary activity in the *in vitro* middle ear mucosa, *Acta Otolaryngol. (Stockholm)*, 104, 495, 1987.
24. **Paparella, M. M., Oda, M., Hiraide, F., and Brady, D.,** Pathology of sensorineural hearing loss in otitis media, *Ann. Otol. Rhinol. Laryngol.*, 81, 632, 1972.
25. **Spandow, O., Anniko, M., and Hellström, S.,** Inner ear disturbances following inoculation of endotoxin into the middle ear, *Acta Otolaryngol. (Stockholm)*, 107, 90, 1989.
26. **Kawauchi, H., DeMaria, T. F., and Lim, D. J.,** The role of the endolymphatic sac in inner ear immune response to middle ear bacterial antigen, *Acta Otolaryngol. (Stockholm)*, 457, 100, 1989.

Chapter 11

DETECTION OF OCULAR INFECTIONS

Eduardo C. Alfonso and Darlene Miller

TABLE OF CONTENTS

I. Introduction 122

II. Keratitis: Current Techniques 122

III. Endophthalmitis: Current Techniques 126

IV. The LAL Test in Ophthalmology 129

V. Future Trends 130
   A. Detection of Endotoxins/Gram-Negative Organisms in Ophthalmic Solutions 130
   B. Detection of Endotoxins/Gram-Negative Organisms in Topical Ocular Medications 131

References 132

I. INTRODUCTION

Ocular structures, especially those at the surface — the conjunctiva and cornea — are constantly exposed to a variety of microorganisms, both pathogenic and nonpathogenic, and yet are seldom infected. This is due in part to an immunologic defense system that embraces local and systemic, specific and nonspecific, and humoral and cellular mechanisms that keep these organisms at bay; it is also due to the anatomic arrangement which leaves the inner ocular structures well protected.[1-3]

Only a few organisms can penetrate the intact mucous lining or epithelium of the conjunctiva and cornea. These include *Neisseria gonorrhoeae*, *Streptococcus pneumoniae*, *Haemophilus influenzae*, *Neisseria meningitidis*, and *Listeria monocytogenes*.[4-10] As long as the epithelium of the external structures remain intact, it forms a protective barrier against invading and opportunist microorganisms.[9,11-13] When this barrier is breached, whether by trauma or by other insults, including insertion and removal of contact lens, it results in a breakdown of these protective ocular structures, leading to invasion by the microorganism.[2,14] An infection that starts in one layer of the eye can rapidly spread to adjacent tissues and cause permanent damage to the functional integrity of the eye.

Since the majority of ocular infections are initiated via external invasion, the structures most frequently infected are the conjunctiva (conjunctivitis), lids (blepharitis), cornea (keratitis), and the skin surrounding the eyes (cellulitis).

Other ocular sites that become infected by contiguous spread are the orbit (orbital cellulitis), the canaliculi (canaliculitis) and the lacrimal sac (dacryocystitis). Infections of the inner chambers of the eye are termed endophthalmitis (Table 1).

Keratitis and endophthalmitis are considered true ocular emergencies.[15] A few hours' delay in diagnosis and treatment can result in severe or permanent ocular damage. This is particularly true when *Pseudomonas aeruginosa*, *Serratia marcescens* or other Gram-negative organisms are the etiologic agents.[15,47]

During the last 11 years, January 1977 through December 1987, the *Limulus* amoebocyte lysate (LAL) assay was used at the Bascom Palmer Eye Institute (BPEI) as an adjunct to Gram stain smears and cultures to screen and diagnose suspected cases of Gram-negative ocular infections. It has proven to be a rapid, reliable, and inexpensive test that can assist in early, specific, and appropriate therapy.[16-18]

Since the LAL test has been used more extensively at the BPEI with the diagnosis and management of keratitis and endophthalmitis, this chapter focuses only on these two clinical entities.

II. KERATITIS: CURRENT TECHNIQUES

Keratitis is an active invasion of the cornea by microorganisms, usually preceded by trauma or other insults. Very few organisms can cross the intact corneal epithelium, but once it is compromised, any organisms can initiate an infection.[4-9,11]

Often bacterial corneal ulcers caused by different organisms look similar but vary in their severity. Ulceration of the cornea in almost all cases is preceded by some form of trauma. Other predisposing features include advancing age, preexisting corneal or external ocular disease, contact lens wear, or instillation of contaminated solutions or medications. Clinically, the hallmark of cornea infections is necrosis.[4,8] The ulcer usually presents as an acute, rapidly progressing, yellow to grayish stromal infiltrate that can conclude in perforation of the cornea and permanent loss of vision within 48 h. Patients complain of a feeling of "sand in the eye", severe pain, sensitivity to light (photophobia) and decreasing vision.[4-9,11] A discharge may or may not be present (Figure 1).

Trauma provides an opportunity for organisms to establish the first step in the pathologic

TABLE 1
Common Bacterial Ocular Isolates, BPEI, 1977—1987[a]

Disease	Isolate
Conjunctivitis	*Staphylococcus aureus*
	Haemophilus influenzae
	Neisseria gonorrhoeae
	Pseudomonas aeruginosa
	Streptococcus pneumoniae
	Proteus mirabilis
Keratitis	*Pseudomonas aeruginosa*
	Staphylococcus aureus
	Fusarium solani
	Serratia marcescens
	Proteus mirabilis
Endophthalmitis	*Staphylococcus epidermidis*
	Staphylococcus aureus
	Streptococcus viridans group
	Group D Streptococcus
	Pseudomonas aeruginosa

[a] Organisms are listed in descending order of occurrence.

process — adhesion. Once adhered, the organisms can enter, multiply, and spread to adjacent tissues. Interaction with the host's inflammatory response exaggerates that tissue damage. If the infection is contained, healing leads to the formation of scar tissue, but if the infection is not controlled, it may lead to loss of the eye.[12,13,19,20]

Proper medical management of bacterial keratitis — in all cases ocular emergencies — must include early identification and elimination of the offending organisms, reduction of the deleterious effects of the inflammatory response and promotion of epithelial healing with minimum scarring.[10,11]

Although treatment is begun as soon as possible when patients present with keratitis, diagnosis and appropriateness of therapy must be confirmed by laboratory studies.[5] Current techniques in ocular microbiology include collection of corneal scrapings for culture (bacterial, viral, fungal, or parasitic) and staining (Gram, Giemsa, acid fast, immunofluorescent, and calcufluor white). Scrapings of the infected cornea are collected with a sterile platinum spatula under the magnification of a slit-lamp or microscope with the use of topical anesthesia. The scrapings are inoculated as collected onto the appropriate media and placed directly on glass slides for smears.

From January 1977 through December 1987, 2102 corneal specimens were collected at the BPEI referral center and submitted for microbiological evaluation. The most frequent isolates are listed on Table 2. Of these, 48% (1014) were culture positive (27% Gram negative, 13% Gram positive, and 8% fungal). Gram-negative organisms remain the predominant group of bacteria isolated in microbial keratitis in South Florida. The number of ulcers in which *P. aeruginosa* has been recovered doubled since the last series presented by Liesegang and Forster,[21] from 31% (1969 to 1977) to 64% (1977 to 1987) (Table 3).

The overall increase from 1963 to 1987 in *Pseudomonas* keratitis was 15%. The last 10 years (1977 to 1987) were the first time in which *P. aeruginosa* has constituted more than 50% of the Gram-negative ulcers. *S. marcescens,* an organism most often involved in opportunistic and nosocomial infections, was a distant second at 9%, followed by *Proteus mirabilis* (6%), *Morganella morganii* (2.6%), and *H. influenzae* (2.4%). *Moraxella* sp., excluding the newly

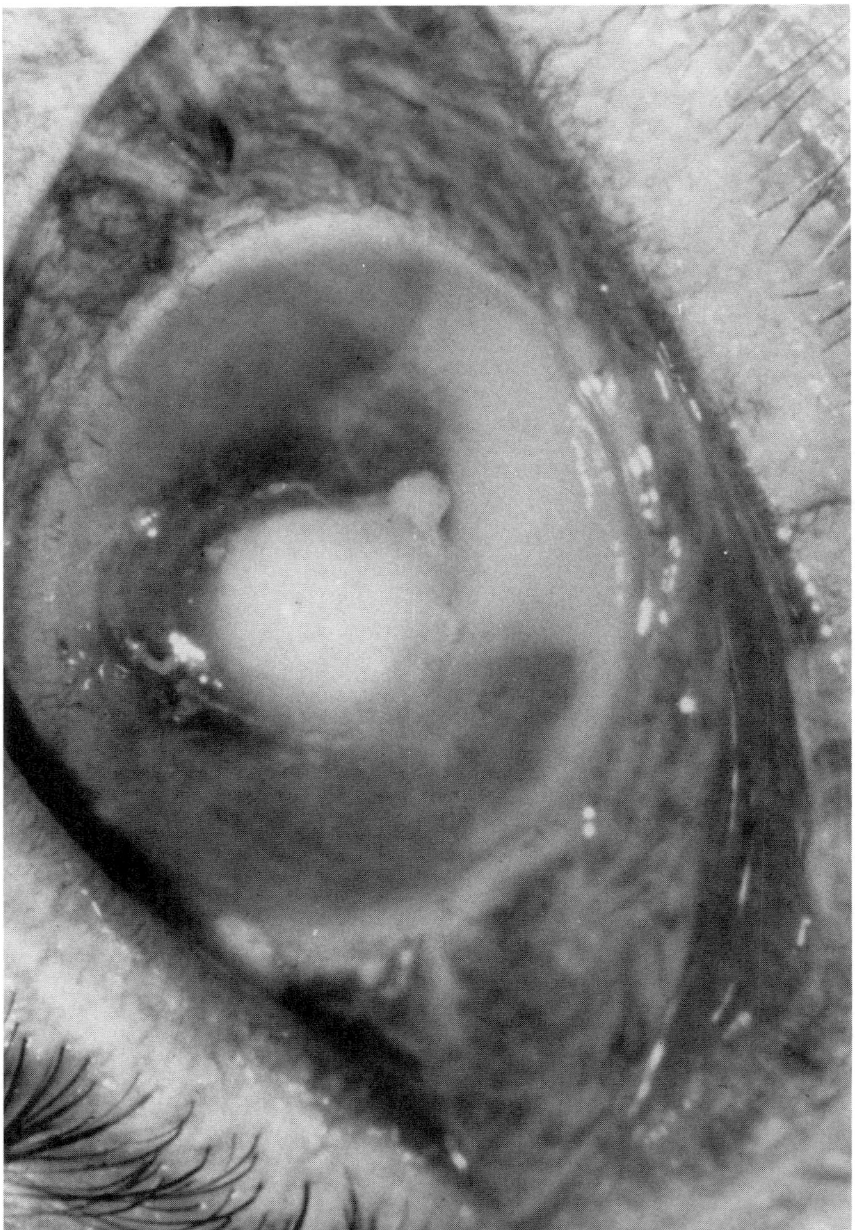

FIGURE 1. Keratitis showing ulceration of the cornea.

TABLE 2
Most Frequent Keratitis Isolates, BPEI, 1977—1987

Isolate	Number
Gram-negative bacteria	
Pseudomonas aeruginosa	367
Serratia marcescens	53
Proteus mirabilis	36
Morganella morganii	15
Haemophilus influenzae	14
Gram-positive bacteria	
Staphylococcus aureus	136
Staphylococcus epidermidis	33
Streptococcus pneumoniae	27
Streptococcus viridans group	20
Corynebacterium species	16
Fungi	
Fusarium sp.	98
Fusarium solani	78
Fusarium oxysporum	17
Fusarium dimerum	2
Fusarium moniliforme	1
Candida sp.	19
Candida albicans	8
Candida parapsilosis	9
Candida tropicalis	1
Candida guillermondi	1
Curvularia sp.	17
Curvularia verruculosa	7
Curvularia senegalensis	7
Curvularia lunata	1
Curvularia sp., NOS	2
Total no. keratitis isolates	1014
Gram-negative bacteria	576
Gram-positive bacteria	276
Total no. fungal isolates	162
Total no. keratitis specimens	2110

TABLE 3
Percent *Pseudomonas aeruginosa* Keratitis Ulcers Vs. Total Number of Gram-Negative Ulcers, BPEI, 1963—1987

	1963—1969	1969—1977	1977—1987
% Gram-negative bacteria	62	49	57
% *P. aeruginosa*	49	31	64

annexed *Moraxella (Branhamella) catarrhalis* constituted 1.4% of the Gram-negative isolates.

Staphylococcus aureus is the most common Gram-positive bacteria involved in this condition; it is responsible for almost half (49.3%) of all Gram-positive keratitis. It is the second most common isolate in cases of infectious keratitis. *S. epidermidis* (11.8%), *S. pneumoniae* (9.7%), *S. viridans* (7.2%), and *Corynebacterium* sp. (6.8%) are the other commonly involved Gram-positive organisms (Table 2).

There has been an increased recovery of *Mycobacterium* sp. other than *Mycobacterium tuberculosis* as causative agents in infectious keratitis at the BPEI. *M. chelonai* (70%), *M. fortuitum* (20%), and *M. avium* complex (10%), are the most common isolates.

Patients on long-term treatment with steroids, antibiotics, or antivirals for ocular disorders are more susceptible to invasion of the cornea by fungi. Trauma, primarily agricultural, is also a frequently elicited history in patients with fungal keratitis. The highest incidence of fungal keratitis is seen in the South and Southeastern U.S., which reflects the subtropical and agricultural nature of the regions.[22]

A chronic, progressive keratitis, increasingly associated with contact lens wear as well as hot tubs and traumatic insult involving soil or fresh water, is caused by the *Acanthamoeba* sp., a freshwater and soil amoeba. This keratitis mimics herpetic and bacterial keratitis. A correct diagnosis is important because a delay in proper treatment will decrease the chances for good visual recovery.[23]

III. ENDOPHTHALMITIS: CURRENT TECHNIQUES

Endophthalmitis, an infection or inflammation involving one or both chambers of the eye, is a catastrophic development following intraocular surgery or penetrating injury. Although still very rare, incidence ranging from 0.07 to 1.0/1400 operations, it remains the most compromising and vision threatening of all ocular infections[11] (Figure 2).

The anterior chamber is able to handle the insult of infectious agents better than the vitreous cavity.[7,8] The fluid of the anterior chamber is clear and resembles the composition of the plasma. The vitreous is clear but gelatinous, made up of 99% water, 1% hyaluronic acid, and collagen molecules. In the face of injury (traumatic or surgical), the anterior chamber more readily allows the infiltration of polymorphonuclear leukocytes (PMNs) and other phagocytic cells. The vitreous, whose gel-like matrix provides a barrier that remains relatively impervious to such cells, provides an excellent culture medium for the growth of the infecting microorganisms.[7,8,15]

Although microbial agents are responsible for the majority of endophthalmitis cases, they may also be initiated by noninfectious agents including excessive surgical manipulation, toxic and chemical substances, immune complexes, implantation of foreign material, and trauma. Infectious endophthalmitis may result as a complication of surgery, or from contiguous spread from adjacent tissues, use of contaminated solutions or medications, endogenous septicemia, and penetrating injuries to the eye.[9,11]

As in keratitis, any organism in sufficient numbers gaining access to either chamber will produce infection. The spectrum of microbial infection may be mild and indolent, presenting several days to weeks after surgery. This may be caused by organisms of low virulence, i.e., coagulase negative staphylococci species and *Propionibacterium acnes*, in contrast to fulminant and invasive endophthalmitis occurring 24 to 48 h after surgery and involving such organisms as *Bacillus cereus* or *P. aeruginosa*. Viral and fungal endophthalmitis are less frequent.[11]

Organisms usually gain entry to the eye at the time of surgery or trauma. The organism is most likely to come from the patient's lid, conjunctiva, or lacrimal sac. Other sources include organisms implanted due to a break in the surgical sterile technique, contaminated donor tissue, medications, or solutions.

The prognosis in microbial endophthalmitis is dependent on several factors. First is the virulence of the invading organisms.[4,11,24-26] When the clinical course is indolent and involves the less pathogenic organisms, and if detected in time, the probability is good that useful vision will be salvaged. However, when organisms such as *Bacillus* sp. or enteric organisms are the causative agents, prognosis is less optimistic.[8] Other factors influencing the visual outcome are

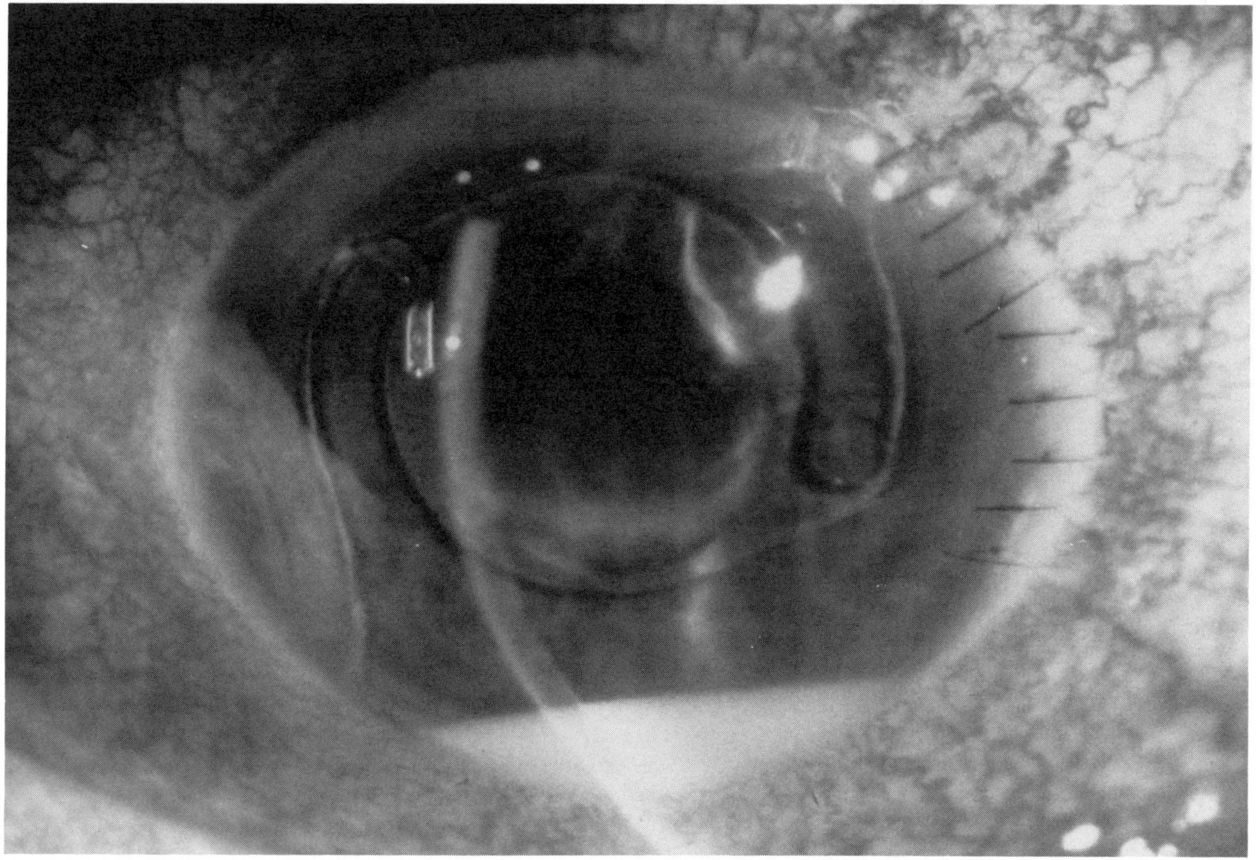

FIGURE 2. Endophthalmitis.

dependent upon an early, rapid, accurate recognition and identification of the disease and organism(s) involved.

Current microbiological techniques for detection of infectious endophthalmitis are the same as in keratitis and include collection of samples for culture and smears. As in keratitis, obtaining appropriate specimens is the hallmark of successful management and treatment of endophthalmitis. Samples for culture and smears are obtained from the anterior and vitreous chamber by the ophthalmologist, using a needle and syringe or mechanized vitrectomy instruments. Vitreous samples are placed directly on the selected media or filtered through a 0.45-mm millipore filter, in which case sections of the filter are placed on the selected media. Organisms are recovered from only 34% of all intraocular fluids submitted for culture. Fluids from the vitreous yield organisms more often than samples from the anterior chamber.[7,8] The ratio at BPEI is 4:1.

Of all microbial endophthalmitis cases, 70% are caused by Gram-positive organisms. Coagulase negative staphylococci species are involved in 30% of all infectious endophthalmitis cases and 43% of all endophthalmitis cases involving Gram-positive organisms. *S. aureus*, *Streptococcus viridans*, and Group D *Streptococcus* and *Bacillus* sp. make up the most frequent organisms in this category. Anaerobic bacteria account for 5% of the Gram-positive isolates, and *P. acnes* is the most common isolate in this group. Gram-negative organisms are implicated in 20% of positive endophthalmitis cases at the BPEI. Not surprisingly, *P. aeruginosa* is the predominant Gram-negative isolate. *H. influenzae* accounts for 4% of the isolates recovered. *S. marcescens* is a frequent Gram-negative isolate and increasingly associated with postoperative instillation of contaminated medications.[8,11]

Fungal endophthalmitis is quite rare and is usually a result of metastasis from an endogenous infectious foci harboring *Candida albicans*. Other species of *Candida* may also be involved. The course is chronic and difficult to manage[9,11] (Table 4).

IV. THE LAL TEST IN OPHTHALMOLOGY

Endotoxins or lipopolysaccharides (LPS) are intricate parts of the cell wall of Gram-negative organisms. They may be found free in the environment during active growth or as remnants of the bacterial cell wall in death.[27-30] Detection of endotoxins in ocular samples can assist the ophthalmologist in selection of therapy and a course of management in cases where growth of the organisms on culture media may be suppressed by pre- or partial treatment with antibiotics, natural inhibitory substances, and inadequacy of sampling.

Over an 11-year period, 554 patients with corneal ulcers (473) or endophthalmitis (81) were screened with the LAL test. This represented 23% of all corneal ulcers and 78% of all endophthalmitis cases submitted for evaluation. Incorporation of the LAL test as an additional screening test has proven it to be a rapid, reliable and easy to perform assay for the detection of Gram-negative endotoxins in ocular tissue and fluids.

The basis of the assay is that even in the presence of minute amounts (picograms) of endotoxins, the lysate from amoebocytes of the horseshoe crab *Limulus polyphemus* or *Tachypleus tridentatus* will form a firm gel-clot that will not disperse upon a 180° inversion of the test vial. In the absence of endotoxins, the lysate remains fluid. The reaction is enzymatic in nature[28,31] (see also Chapter 3).

The *Limulus* lysate, which was obtained commercially from Associates of Cape Cod, Woods Hole, MA, was freeze-dried in single-test vials with a predetermined sensitivity of 0.125 ng/ml. Controls, both positive and negative, were set up with each testing of patient samples. The positive control (purified endotoxin from *Escherichia coli*) was also commercially obtained. Negative controls consisted of the lysate plus 0.2 ml of pyrogen-free water. All patient samples were collected by an ophthalmologist and sent to the microbiology laboratory for evaluation. Corneal scrapings were collected under topical anesthesia, using a platinum spatula. The

TABLE 4
Most Frequent Endophthalmitis Isolates, BPEI, 1977—1987

Isolate	No.
Gram-negative bacteria	
Pseudomonas aeruginosa	14
Serratia marcescens	13
Proteus mirabilis	11
Morganella morganii	8
Citrobacter diversus	6
Gram-positive bacteria	
Staphylococcus sp. (coagulase negative species)	110
Staphylococcus epidermidis	106
Staphylococcus warneri	2
Staphylococcus capitis	1
Staphylococcus hominis	1
Staphylococcus aureus (coagulase positive)	46
Streptococcus viridans group	37
Streptococcus viridans, NOS	24
Sttreptococcus sanguis	6
Streptococcus morbillorum	5
Streptococcus mitis	1
Streptococcus MG-intermidius	1
Group D-*Streptococcus* group	16
Streptococcus faecalis (Enterococcus faecalis)	15
Streptococcus bovis	1
Streptococcus pneumoniae	10
Fungi	
Candida species	12
Candida albicans	9
Candida parapsilosis	2
Candida tropicalis	1
Fusarium species	8
Fusarium solani	7
Fusarium oxysporum	1
Aspergillus species	4
Aspergillus flavus	3
Aspergillus fumigatus	1
Total no. Gram-positive bacteria	276
Total no. Gram-negative bacteria	72
Total no. fungal isolates	4
Total no. endophthalmitis isolates	364

collected materials were then rinsed into a 2-ml vial containing pyrogen-free water, labeled, and sent to the laboratory for processing. Intraocular fluids were collected by the ophthalmologist at the time of surgery and at least 0.3 ml of sample inoculated into a pyrogen-free tube, labeled, and sent to the laboratory for processing.

The LAL test was performed by adding 0.2 ml of patient samples to each single-test vial and 0.2 ml of pyrogen-free water to each control vial. Once rehydrated and mixed, all LAL test vials (patient and controls) were incubated undisturbed at 37°C for 1 h. Following incubation, controls were examined for expected results: positive, indicated by formation of a firm gel-clot that did not dissolve upon 180°C inversion, or negative, indicated by formation of a soft clot that

TABLE 5
Sensitivity, Specificity, and Overall Accuracy of the LAL Test to Predict Gram-Negative Culture Results from Keratitis and Endophthalmitis Specimens, BPEI, January 1977—December 1987

		LAL test results (%)		
Disease	No. of specimens	Sensitivity	Specificity	Overall accuracy
Keratitis	473	89.7	93.4	91.5
Endophthalmitis	81	90.0	93.4	92.6

readily dispersed upon inversion or no clot formation. After controls were recorded, patient samples were evaluated and scored according to the above guidelines.[28]

Demonstration of organisms in culture continues to be the standard by which all new tests must be judged. The specificity, sensitivity, and predictive value of the LAL test was evaluated in cases of keratitis and endophthalmitis referred to the BPEI (Table 5). The number of false-positive cases is due primarily to the detection by the LAL test of endotoxins in the absence of growth in culture.[28,30-32] Of patients seen at our referral center, 95% have some history of antibiotic therapy prior to obtaining a culture. The treatment is selected to eliminate both Gram-positive and Gram-negative flora; therefore, patients presenting with a Gram-negative infection may have a positive LAL test in the absence of growth of the organism in culture. In other patients, Gram-negative organisms may be observed in Gram stain, and fail to grow in culture, yet still produce a positive LAL test. Contamination of the LAL test with Gram-negative organisms could lead to a false-positive result. At the opposite end of the spectrum and perhaps the more important are the false-negative results. Causes of false-negative results include failure to inoculate the collection vial for the LAL test, the presence of inhibitory substances, and selection of a lysate below the threshold of sensitivity for detection of endotoxins.[30]

The LAL test is a rapid, accurate, reliable, and easy to perform test for the detection of endotoxins in corneal scrapings or intraocular fluids. Results obtained can help guide the physician in the selection and initiation of an early, specific and appropriate therapy. In those cases where the patient result might be influenced by pre- or partial antibiotic treatment, it can offer a clue as to the etiologic agents involved.

Table 5 provides a summary of 11 years of experience with the LAL test at the BPEI. When the LAL test was positive (sensitivity), 89.7% of ocular samples from keratitis and 90.0% of samples from endophthalmitis grew Gram-negative organisms in culture. When the LAL test was negative (specificity), 93.4% of ocular samples from both keratitis and endophthalmitis did not grow Gram-negative organisms in culture. The overall accuracy of the LAL test in predicting Gram-negative culture results was 91.5 and 92.6% for keratitis and endophthalmitis, respectively.

V. FUTURE TRENDS

A. DETECTION OF ENDOTOXINS/GRAM-NEGATIVE ORGANISMS IN OPHTHALMIC SOLUTIONS

Contamination of contact lenses and their cleaning solutions with Gram-negative bacteria — in particular *P. aeruginosa* and S. *marcescens* — and their potential to cause ocular infections is a growing concern to ophthalmologists.[33-38] Although the danger of microbial keratitis in contact lens wearers is low, many eyes are at risk, as at least 18 million people have been fitted with contact lenses since 1983.[35] This keratitis can lead to permanent visual impairment in an otherwise healthy eye. Among the leading risk factors for incurring this type of infection is poor patient hygiene, which can lead to microbial contamination of the lenses and their solutions.[30,33,34] Soft extended wear lenses are more often associated with Gram-negative keratitis than are the hard lenses. The use of the LAL test for the early and rapid detection of *P. aeruginosa* and *S. marcescens* contamination of contact lenses and their solutions could reduce the risk of developing the severe, often vision-impairing keratitis associated with these organisms.

The use of the LAL test for detection of contamination of contact lens solutions was studied at the BPEI. Using multiple aliquots from 17 unopened brands (15 for soft, 2 for hard), plus multiple aliquots from these brands mixed with endotoxin, the ability of two LAL products (#1 — [Pyrotell] Associates of Cape Cod, #2 — [Pyrogent] Whittaker Co.) with a sensitivity of 0.125 endotoxin units (EU)/ml to detect the presence or absence of endotoxins in commercially available contact lens solutions was evaluated. Sensitivity was 65% (11 of 17) for LAL #1 and 85% (14 of 17) for LAL #2. Samples not forming gel-clots were found to have inhibitory products that when diluted 1:4 or higher with pyrogen-free water increased the sensitivity to 99%. The solutions used for hard contact lenses had the highest amount of inhibitory substances. Utilization of the LAL test with the appropriate controls and dilutions for the detection and monitoring of contaminated solutions may offer an early, rapid, and reliable warning system for reducing and avoiding the ulcerative keratitis associated with contact lens wear.[29,40]

B. DETECTION OF ENDOTOXINS/GRAM-NEGATIVE ORGANISMS IN TOPICAL OCULAR MEDICATIONS

Despite the fact that topical medications in the U.S. contain preservatives, either at bacteriostatic or bactericidal levels, reports of keratitis, endophthalmitis, sclerokeratitis, and conjunctivitis occurring in patients using contaminated topical medications are appearing with increasing frequency.[36,38] Gram-negative organisms, especially *P. aeruginosa* and *S. marcescens* are among the most frequent isolates. The organisms can survive and multiply in many disinfectants, including hexachlorophene and benzalkonium chloride, a common eye drop preservative.[41,42]

Organisms may colonize and multiply in ocular medications through repeated inoculation from the patient's hands, lids, or ocular surface.[36-38,41] Once the inhibition threshold of the preservatives has been overcome, the organisms can grow and be dispensed in sufficient numbers to establish infection. The organisms may also adhere to the outer part of the tip of the dispensers or to the inner part of the bottle cap. One mechanism for overcoming the bacteriocidal or bacteriostatic threshold of the preservative is the formation of a biofilm that protects the organism from the action of preservatives.

A total of 228 topical medications were collected from patients seen at our hospital from 1977 to 1987; screening revealed a contamination rate of 23%. Of these isolates, 12% were also isolated from cultures of these patients. Culture isolates included keratitis (82%), endophthalmitis (11%), and scleritis (7%). In each case, bacteria of identical biotype and sensitivity were isolated from the affected eye and at least one of the patient's medications. Gram-negative isolates comprised 84% of the total recovered.

Preliminary testing at the BPEI suggests that with appropriate controls, proper pH adjustments, and two- to fourfold dilutions, the LAL test should be investigated as a screening assay for detecting Gram-negative contamination of ocular medications.[43-46]

REFERENCES

1. **Thoft, R. A. and Friend, J.,** *Int. Ophthalmol. Clin.*, 2, 111, 1979.
2. **Allansmith, M. R.,** Defense of the ocular surface. *Int. Ophthalmol. Clin.*, 2, 93, 1979.
3. **Novitsky, T. J.,** LAL water, in *LAL Update*, Vol. 3, No. 5, Associates of Cape Cod, Inc., Woods Hole, MA, September 1985.
4. **Wilson, L., Ed.,** *External Disease of the Eye*, Harper & Row, New York, 1979, chap. 5—10.
5. **Bohigan, G. M.,** *Handbook of External Diseases of the Eye*, Slack, Thorofare, NJ, 1987, chap. 1—17.
6. **Fedukowicz, H. B. and Stenson, S.,** External infections of the eye, in *Bacterial, Viral Myotic with Noninfectious and Immunologic Diseases*, Appleton-Century-Crofts, Norwalk, CT, 1985, chap. 1—5.
7. **Thomas, D., Ed.,** *Clinical Ophthalmology, External Disease, the Uvea*, Vol. 4, Harper & Row, New York, 1979, chap. 24.
8. **Tabbara, K. F. and Hyndwk, R. A.,** *Infections of the Eye*, Little, Brown, Boston, 1986, chap. 35.
9. **Locatcher-Khoraza, D. and Segal, B. D.,** *Microbiology of the Eye*, C.V. Mosby, St. Louis, 1972, chap. 1—7, 13.
10. **Friedman, A. H.,** Ocular Infectious Disease, A Guide for the General Practitioner, Seminar Series #7, Abbott Laboratories, North Chicago, IL, 1978.
11. **Smolin G., Tabbara, K., and Whitcher, J.,** *Infectious Diseases of the Eye*, Williams & Wilkins, Baltimore, 1984.
12. **Ramphas, R., McNiece, M. T., and Polack, F. M.,** Adherence of *Pseudomonas aeruginosa* to the injured cornea. A step in the pathogenesis of corneal infections, *Ann. Ophthalmol.*, 13, 421, 1981.
13. **Das, N. D., Zam, Z. S., and Reichert, R.,** Adherence properties of *Pseudomonas* pili to epithelial cells of human cornea, additional "lectionological" aspects, *Curr. Res.*, 2, 641, 1983.
14. **Chandler, J. W. and Tilette, T. E.,** Immunologic defense mechanisms of the ocular surface, *Ophthalmology*, 90(6), 585, 1983.
15. **Vaughn, D. and Asbury, T.,** *General Ophthalmology*, 11th ed., Lange Medical, Appleton and Lange, Los Altos, CA, 1986.
16. **Miller, D. and Forster, R. K.,** Evaluation of the Limulus Lysate Assay as a Rapid Screening Method for Detection of Endotoxins in Microbial Keratitis and Endophthalmitis, Abstract #B178, Annual Meeting of the American Society for Microbiology, March 23 to 28, 1986, Washington, D.C.
17. **Miller, D. and Forster, R. K.,** Utilization of the LAL as a rapid screening method for detection of endotoxins in microbial keratitis. An eight year study, *Invest. Ophthomol. Vis. Sci.*, 27, 36, 1986.
18. **McBeath, J., Forster, R. K., and Rebell, G.,** Diagnostic limulus lysate assay for endophthalmitis and keratitis, *Arch. Ophthalmol.*, 96, 1265, 1978.
19. **Gristina, A. G.,** Biomaterial-centered infection, microbial adhesion versus tissue integration, *Science*, 237, 1588, 1987.
20. **Savage, D. C. and Fletcher, M. M.,** *Bacterial adhesion, Mechanism and Physiological Significance*, Plenum Press, New York, 1985, chap. 1—5.
21. **Liesegang, T. J. and Forster, R. K.,** Spectrum of microbial keratitis in South Florida, *Am. J. Ophthalmol.*, 90, 38, 1980.
22. **Asbell, P. and Stenson, S.,** Ulcerative keratitis — survey of 30 years laboratory experience, *Arch. Ophthalmol.*, 100, 77, 1982.
23. **Jones, D. B.,** Acanthamoeba — the ultimate opportunist?, *Am. J. Ophthalmol.*, 102(4), 527, 1986.
24. **Forster, R. K., Abbot, R. L., and Gelender, H.,** Management of infectious endophthalmitis, *Ophthalmology*, 87, 313, 1980.
25. **Forster, R. K.,** Etiology and diagnosis of bacterial postoperative endophthalmitis — symposium, postoperative endophthalmitis, *Ophthalmology*, 85, 320, 1978.
26. **Forster, R. K.,** Endophthalmitis, diagnostic cultures and visual results, *Arch. Ophthalmol.*, 92, 387, 1974.
27. **Landy, M. and Baum, W.,** *Bacterial Endotoxins, Proceedings of a Symposium Held at the Institute of Microbiology of Rutgers, the State University*, Quinn & Boder, Richway, NJ, 1964.
28. **Novitsky, T. J.,** Discovery to commercialization — the blood of the horseshoe crab, *Oceanus*, Vol. 27, Spring 1984.
29. **Novitsky, T. J.,** International conference of endotoxin assay (ICEA) implications for medicine and production quality control, in *LAL Update*, Vol. 2, No. 4, Associates of Cape Cod, Inc., Woods Hole, MA, July 1984.
30. **Dawson, M. E., Novitsky, T. J., and Gould, M. J.,** Microbes, endotoxins and water, *Pharm. Eng.*, 8, 2, 1988.
31. **Gould, M. C.,** Endotoxin in Vertebrate Cell Culture, Its Measurement and Significance, Monograph #5, Tissue Cultures Association, Inc., Woods Hole, MA, 1984.
32. **Novitsky, T. J.,** The biology of the horseshoe crab, in *LAL Update*, Vol. 3, No. 4, Associates of Cape Cod, Inc., Woods Hole, MA, July 1985.
33. **Parment, P. and Ronerstam, R. A.,** Soft contact lens keratitis associated with *Serratia marcescens*, *Acta Ophthalmol.*, 59, 560, 1981.

34. **Chalupa, E., Swarbrick Holden, B. A., and Sjostrand, J.,** Severe corneal infection associated with contact lens wear, *Ophthalmology*, 94, 17, 1987.
35. **Alfonso, E., Mandelbaum, S., Fox, M. J., and Forster, R. K.,** Ulcerative keratitis associated with contact lens wear, *Am. J. Ophthalmol.*, 101, 429, 1986.
36. **Templeton, W. C., III, Eiferman, R. A., Snyder, J. W., Melo, J. C., and Raff, M. J.,** Serratia keratitis transmitted by contaminated eyedroppers, *Am. J. Ophthalmol.*, 93, 723, 1982.
37. **Saleda, S. R., Lapuz, J., and Vizcone, R.,** Serratia marcescens endophthalmitis, *Arch. Ophthalmol.*, 89, 163, 1973.
38. **Cuad, C. T., Osata, M. S., and Wilhelmus, K. R.,** Bacterial contamination of eye drop dispenses, *Am. J. Ophthalmol.*, 98, 548, 1984.
39. **Alfonso, E., Miller, D., and Forster, R. K.,** Comparative study of two LAL reagents for detection and monitoring of contaminated contact lenses solutions for the presence of endotoxins, *Invest. Ophthomol. Vis. Sci. Suppl.*, 3, 501, 1989.
40. **Alfonso, E., Miller, D., and Forster, R. K.,** Use of Limulus Amoebocyte (LAL) in Detection and Monitoring of Contaminated Contact Lenses and Solutions for the Presence of Gram-negative Organisms (Endotoxins), Abstract #B291, Annual Meeting of the American Society for Microbiology, May 14 to 19, 1989, New Orleans.
41. **Schein, O. D., Wasson, P. J., Boruchoff, S. A., and Kenyon, K. R.,** Microbial keratitis associated with contaminated ocular medications, *Am. J. Ophthalmol.*, 105, 361, 1988.
42. **Vaughn, D. G., Jr.,** The contamination of fluorescein solutions with special reference to *Pseudomonas aeruginosa (bacillus pyocyaneos)*, *Am. J. Ophthalmol.*, 39, 55, 1955.
43. **Miller, D., Holland, S. P., Song, D., Alfonso, E., Costerton, J. W., and Ruseska, I.,** Role of Biofilm in Persistent *Pseudomonas aeruginosa* Ocular Infections, Abstract #B242, Annual Meeting of the American Society for Microbiology, May 14 to 19, 1989, New Orleans.
44. **Miller, D., Hollands, S. P., Song, D., Alfonso, E., Costerton, J. W., and Ruseska, I.,** Role of glycocalyx in recurrent and persistent *Pseudomonas aeruginosa* ocular infections, *Invest. Ophthomol. Vis. Sci. Suppl.*, 3, 196, 1989.
45. **Holland, S. P., Miller, D., Chuang, E., Childress, L., Rosenwasser, G., Ruseska, I., and Costerton, J. W.,** Biofilm, a method of bacterial survival in contaminated topical medications, *Invest. Ophthamol. Vis. Sci. Suppl.*, 3, 503, 1989.
46. **Costerton, J. W., Lam, J., and Chan, R.,** The role of the microcolony mode of growth in the pathogenesis of *Pseudomonas aeruginosa* infections, *Rev. Infect. Dis.*, 5(Suppl.), 5867, 1983.
47. **Alfonso, E., Kenyon, K., Ormerod, L. D., Stevens, R., Wagoner, M. D., and Albert, D. M.,** Pseudomonas corneoscleritis, *Am. J. Ophthalmol.*, 103, 90, 1987.

Chapter 12

DETECTION OF PERITONITIS AND PERFORATION IN BLUNT ABDOMINAL TRAUMA

David L. Smalley

TABLE OF CONTENTS

I. Peritonitis in Continuous Ambulatory Peritoneal Dialysis 136
A. Background 136
B. *Limulus* Amoebocyte Lysate Test 137

II. Perforation in Blunt Abdominal Trauma 137

References 138

I. PERITONITIS IN CONTINUOUS AMBULATORY PERITONEAL DIALYSIS

A. BACKGROUND

Continuous ambulatory peritoneal dialysis (CAPD) has become a renewed treatment for end-stage renal disease. Currently, there are over 20,000 end-stage renal disease patients maintained on CAPD worldwide, with the numbers increasing dramatically each year.[1] In comparison with hemodialysis, CAPD offers several advantages: (1) the individual patient can be taught how to manage the fluid exchange process at home; (2) the risk of infectious diseases such as hepatitis B is markedly reduced; and (3) the cost of CAPD is only a fraction of the cost of hemodialysis.

The principle complication associated with the use of CAPD is bacterial peritonitis. Obviously, the possibility of contamination while performing the fluid exchange is quite high unless extreme precautions are taken. Typically, patients are trained thoroughly in aseptic technique but most can still expect to develop peritonitis at some point.

Gram-positive cocci have been associated with over 60% of the cases reported, with *Staphylococcus epidermidis* identified as the prevailing organism.[2] In other studies, Gram-negative organisms account for 15 to 25% of the bacterial peritonitis cases among CAPD patients[3] while 4 to 28% were culture negative.

The fluid from a CAPD exchange is normally crystal clear, therefore, the typical case of bacterial peritonitis begins with the patient developing a cloudy dialysate noted during a fluid exchange. The cloudiness is a result of inflammatory response and the presence of polymorphonuclear leukocytes (PMNs).[4] Following the observation of cloudy fluid, the patient may begin to develop abdominal pain, tenderness, and nausea. The CAPD patients are usually afebrile during a bacterial peritonitis event, probably due to the volume of fluid exchange rate and the rare development of bacteremia.

The nephrologist will usually have the cloudy fluid undergo laboratory evaluation, including a white blood cell (WBC) count, leukocyte differential, and culture. Culture of the CAPD dialysate may be performed in several different ways, including special centrifugation and filtration prior to culture.[3,4] The typical bacterial peritonitis case has fluid containing PMNs of greater than 100/mm^3.[5] Recent studies of these fluids revealed that they can easily be screened for an inflammatory response using the leukocyte esterase test (LET).[6-8] The LET was developed to detect PMNs in urine at the level of ten cells per cubic millimeter.[9] However, due to the hypertonicity of the CAPD dialysate, the LET sensitivity is reduced to detect 100 PMNs per cubic millimeter or greater. Fortunately, this correlates with the clinically defined amount of inflammatory response necessary to declare clinical significance associated with bacterial peritonitis. Presently, the gross observation of cloudy fluid, confirmation of inflammatory response, and clinical presentation of physical findings previously described directs the nephrologist to the diagnosis of bacterial peritonitis.

Among the tests used for diagnosis, the culture of the CAPD dialysate is the most significant. Because the majority of the bacterial peritonitis cases are caused by coagulase negative staphylococci, empiric therapy with cephalosporins is the most common treatment. Due to the clinical similarities of Gram-positive and Gram-negative peritonitis, one may expect to successfully treat a majority of the cases empirically. However, Gram-negative organisms may be resistant to the cephalosporins most commonly used, and treatment with these agents may not result in improvement and resolution of abdominal pain, tenderness, or nausea. The patient will be treated next with an aminoglycoside or combination of an aminoglycoside and cephalosporin. Unfortunately, the delay of 24 to 48 h in culture information leaves the nephrologist to treat only on the basis of clinical judgment. It has also been recognized that the more severe the case of peritonitis and the longer it persists, the greater the likelihood of tissue damage of the peritoneal wall, thus creating future problems in exchange by CAPD.

TABLE 1
Results of LAL Tests on 185 Specimens Compared to Results of Cultures

	LAL assay results[a]	
Culture result	**No. positive**	**No. negative**
Gram negative	66	0
Gram positive	9	40
Fungus positive	0	7
Culture negative	5	58

[a] Sensitivity for LAL assay in detecting Gram-negative peritonitis was 100%; specificity was 88.2%.

B. *LIMULUS* AMOEBOCYTE LYSATE TEST

In order to expedite the diagnosis of a Gram-negative infection, several studies have been undertaken to evaluate the use of the *Limulus* amoebocyte lysate (LAL) test in correlation to culture. In one study reported,[10] the micromethod,[11] LAL gelation procedures were used. Among ten culture-proven cases of Gram-negative peritonitis, seven were positive for endotoxin by the LAL test. In addition, only 1 of 33 culture-proven cases of Gram-positive cocci had positive results by the LAL test. Gram-negative organisms causing these 10 infections included *Pseudomonas alcaligenes* (3), *Enterobacter cloacae* (2), *E. cloacae* and *P. maltophilia* (1), *Klebsiella pneumoniae* plus *Bacteroides fragilis* (1), *Citrobacter freundii* (1), and Centers for Disease Control VEI (1).

In studies performed in our laboratories, the chromogenic LAL test was used to test 66 culture-proven Gram-negative peritonitis cases among 185 CAPD patients with clinical evidence of infectious peritonitis.[12] Of of 185 fluids evaluated, all 66 culture-proven Gram-negative infections were LAL positive. Sixty-three specimens were from cases of culture-negative peritonitis, five of which were positive for endotoxins as determined by the LAL test, suggesting that the LAL test may be detecting low-level Gram-negative infections below the sensitivity of present culture methods yielding negative data (Table 1). The chromogenic LAL results on the Gram-negative cases had positive values ≥0.8 endotoxin units (EU)/ml. Based on these data, the sensitivity of the LAL and the predictive value of a negative finding were 100%. The calculated specificity and predictive value of a positive outcome of the LAL test were 88.2 and 82.5%, respectively.

Because CAPD usage is increasing,[1] the future of LAL testing among these patients is encouraging. Peritonitis caused by Gram-negative bacilli is usually more severe clinically than peritonitis caused by Gram-positive cocci.[4] Since there is variance in antimicrobial therapy from peritonitis caused by Gram-negative vs. Gram-positive organisms, the rapid detection of Gram-negative associated endotoxin by the LAL test may aid in early initiation of appropriate antibiotic therapy. With the development of even shorter and more sensitive chromogenic assays, the LAL may become a bedside test or even a home test used by CAPD patients in a manner similar to that used for the LET.

II. PERFORATION IN BLUNT ABDOMINAL TRAUMA

A number of studies have been reported in which the LAL test was utilized to detect endotoxemia associated with different diseases and injuries of the gastrointestinal (GI) tract. In one study,[13] in which the LAL test was used in 274 patients, circulating endotoxin was found in all patients with fulminant hepatic failure and alcoholic cirrhosis. In addition, a varying reaction was found in patients with viral hepatitis, acute pancreatitis, Crohn's syndrome, and ulcerative colitis.

In another study,[14] the LAL test was used to test tissue homogenates of 20 patients. Endotoxemia among these patients was associated with an increased incidence of gastrointestinal bleed and major injury to the liver. The authors suggested that the endotoxemia commonly develops when the antibacterial defense has been depleted due to immunocompromised states.

Although these studies suggest LAL positivity associated with trauma or GI perforation, the preliminary study reported by Tu et al.[15] looked directly at the presence of endotoxin in peritoneal lavage fluid. In their study, the LAL test was compared to Gram stains, acridine orange stains, and cultures to determine whether GI perforation occurred following blunt abdominal trauma. The study looked at 23 patients following colon surgery; 6 patients had lavage fluids that had positive bacterial cultures, and only 3 of the 6 had positive Gram stains or acridine orange stains. The LAL test, using the 1 h gelation procedure, successfully detected all 6 cases, whereas all culture-negative cases revealed negative LAL tests on peritoneal lavages.

With the advent of the chromogenic LAL test, which has a reduced assay time and increased sensitivity over gelation, reevaluation of blunt abdominal trauma associated endotoxin may prove the LAL test to be worthwhile in the detection of GI perforation. In addition, the presence of circulating endotoxins may reveal perforation following certain diseases or infections of the GI tract.

REFERENCES

1. **Rubin, J., Rogers, W. A., Taylor, H. M., Everett, D., Prowant, B. F., Fruto, L. V., and Nolph, K. D.,** Peritonitis during continuous ambulatory peritoneal dialysis, *Ann. Intern. Med.,* 92, 7, 1980.
2. **Gruer, L. D., Bartlett, R., and Ayliffee, G. A. J.,** Species identification and antibiotic sensitivity of coagulase-negative staphylococci from CAPD patients, *J. Antimicrob. Chemother.,* 13, 577, 1984.
3. **Rubin, S. J.,** Continuous ambulatory peritoneal dialysis: dialysate fluid cultures, *Clin. Microbiol. Newsl.,* 6, 3, 1984.
4. **Vas, S. I.,** Microbiologic aspects of chronic ambulatory peritoneal dialysis, *Kidney Int..* 23, 83, 1983.
5. **Duwe, A. K., Vas, S. I., and Weatherhead, J. W.,** Effects of the composition of peritoneal dialysis fluid on chemiluminescence, phagocytosis, and bactericidal activity *in vitro, Infect. Immun.,* 33 130, 1981.
6. **Chan, L. K. and Oliver, D. O.,** Simple method for early detection of peritonitis in patients on continuous ambulatory peritoneal dialysis, *Lancet,* 2, 1336, 1979.
7. **Smalley, D. L., Doyle, V. R., and Duckworth, J. K.,** Correlation of leukocyte esterase detection and the presence of leukocytes in body fluids, *Am. J. Med. Technol.,* 48, 135, 1982.
8. **Smalley, D. L., Kraus, A. P., Jr., and Baddour, L. M.,** Clinical use of the leukocyte esterase test in continuous ambulatory peritoneal dialysis, *Lab. Med.,* 19, 164, 1988.
9. **Smalley, D. L. and Bryan, J. A.,** Comparative evaluation of biochemical and microscopic urinalysis, *Am. J. Med. Technol.,* 49, 237, 1983.
10. **Cartmill, T. D. I., Tapson, J. S., and Freeman, R.,** Limulus lysate study of effluents from CAPD patients, *Peritoneal Dial. Bull.,* 7, 197, 1987.
11. **Flowers, D. J.,** A micro technique for endotoxin assay by using Limulus lysate, *Med. Lab. Sci.,* 36, 171, 1979.
12. **Smalley, D. L., Baddour, L. M., and Kraus, A. P., Jr.,** Rapid detection of Gram-negative bacterial peritonitis by the Limulus amoebocyte lysate assay, *J. Clin. Microbiol.,* 24, 882, 1986.
13. **Liehr, H.,** The Limulus assay for endotoxemia, as applied in gastroenterology, in *Biomedical Applications of the Horseshoe Crab (Limulide),* Cohen, E., Ed., Alan R. Liss, New York, 1979, 309.
14. **Woodruff, P. W. H., O'Carroll, D. I., Koizumi, S., and Fine, J.,** Role of the intestinal flora in major trauma, *J. Infect. Dis.,* 128(Suppl.), 5290, 1973.
15. **Tu, K. K., Kuminsky, R. E., and Boland, J. P.,** *Limulus* amoebocyte lysate test for rapid detection of GI perforation in blunt abdominal trauma, *ASM Abstr.,* C274, 1981.

Chapter 13

ALLIED MEDICAL APPLICATIONS OF THE *LIMULUS* AMOEBOCYTE LYSATE TEST

Frederick C. Pearson

TABLE OF CONTENTS

I. Introduction 140

II. Evaluation of Hemodialyzers and Reuse Water 140
- A. Treatment 140
- B. Pyrogenic Reactions in Hemodialysis 141
- C. Membrane Permeation of Endotoxin 141
- D. Generation of Interleukin-1 by Endotoxin in Hemodialysis 142
- E. Interleukin-1 142
- F. Pyrogen Evaluation of Water in Hemodialysis 144
- G. Water Quality Standards for Hemodialysis Reuse 144
- H. Dialysate 145
- I. Endotoxin Evaluation of Hemodialyzer 146

III. Surveillance of Respiratory Equipment 148

IV. Detection of Contaminated Foods and Water 149

V. Conclusions 150

References 150

I. INTRODUCTION

Hemodialysis (HD) is a therapeutic modality used to purify the blood of those individuals having a build up of uremic toxins and fluid caused by acute or chronic renal failure. Nearly 100,000 individuals suffer from chronic end-stage renal disease in the U.S. alone, and there are over twice this number worldwide, with the primary incidence in Europe and Japan.[1] Hemodialysis patients sometimes experience pyrogenic reaction caused by endotoxin contamination of the water used in several stages of the procedure. The *Limulus* amoebocyte lysate (LAL) test has been developed to detect the presence of endotoxin, and through the conscientious use of this test it is possible to eliminate many of the sources of these pyrogenic reactions.

II. EVALUATION OF HEMODIALYZERS AND REUSE WATER

A. TREATMENT

Vascular access is required to perform hemodialysis, and it is typically accomplished by twin surgical cutdowns over an artery and a vein at the wrist. During extracorporeal hemodialysis, arterial and venous lines are attached to the respective ports on the artificial kidney (AK). The AK is composed of a dialysis membrane package and a fluid cycler. The hemodialysis membrane material is typically composed of cellulose or cellulose-derived material, but synthetic membrane materials are being used more widely today. Membranes are manufactured in a number of formats, including a coil, sheet, or hollow fiber to provide a large contact surface for toxin-containing blood. The patient's blood is forced through the artificial kidney at a conventional flow rate of 200 ml/min. Dialysate, a hyperosmolar solution of electrolytes and a small molecular weight carbon compound, is pumped through the outside of the membrane and causes the flow of water and uremic toxins from the blood through the membrane and ultimately to the dialysate side where the waste products and excess water are pumped away. The most common uremic wastes include urea and creatinine. Hemodialysis is usually considered for patients in whom blood urea nitrogen (BUN) concentration reaches 100 mg/dl or the plasma creatinine level reaches 10 mg/dl. The broad goal of treatment is to hold the BUN and creatinine levels at or below these levels while providing reasonable protein intake and controlling fluid overload. Conventionally, renal failure patients undergo hemodialysis treatment 3 times per week for a period of 4 to 6 h. Most hemodialysis treatments are conducted in centers dedicated to this procedure, but home hemodialysis is more attractive to some patients.

Hemodialysis patients are at risk for a number of maladies, including acute and sometimes profound episodes of pyrogenic reactions due to endotoxin, the outer membrane of Gram-negative bacteria.[2,3] The etiology of these pyrogenic episodes emanates from endotoxin contamination of the water used in the procedure, the dialysate, or the reused hemodialysis membrane. First-use hemodialysis membranes do not have a history of provoking such episodes. In contrast, reused hemodialyzers have been shown to be more biocompatible than new ones with respect to pronounced hypersensitivity reaction (removal of ethylene oxide or other sensitizing agents), complement activation, and other physiological dysfunctions.[4-7] In addition, inadequate preparation of artificial kidneys has been associated with acute episodes of pyrogenicity. Water is a common source of Gram-negative bacteria (and thus, endotoxin) and it must be of a high quality for use in dialysis. While the healthy individual has a weekly intake of about 15 l of drinking water, a patient on hemodialysis treatment is exposed to between 300 and 600 l of treated water per week.[8] This enormous weekly exposure to water mandates that strict water quality standards be met for use in hemodialysis therapy. Acute attacks of pyrogenicity have been associated with contaminated dialysate and are discussed in this chapter. Silicone and polyvinyl chloride tubing used in dialysis have not been implicated in pyrogenic reactions.

An alternative to hemodialysis is continuous ambulatory peritoneal dialysis (CAPD) which, as the name implies, is accomplished by taking advantage of the dialysis capacity of the peritoneal membrane. This procedure can be done by the patient in the home or at work, avoiding the need for in-center hemodialysis treatments. Although this approach requires 80 to 100 l of dialysate to be exchanged by the patient each week, this method of dialysis is not considered here since dialysate solutions, when generated under Good Manufacturing Practices (GMP), have not been linked to pyrogenicity or peritonitis. Typically, subpyrogenic levels of endotoxin have been observed in dialysate prepared for CAPD as well as for hemodialysate[9] (see also Chapter 12).

B. PYROGENIC REACTIONS IN HEMODIALYSIS

Despite technical improvements in hemodialysis over the past several decades (including the introduction of disposable equipment), patients undergoing chronic hemodialysis are subject to considerable risks, including acute episodes of fever and infection. Although fever is a common manifestation of infection, a significant number of febrile reactions observed during hemodialysis are not caused by demonstrable culture-proven infection.[10] While the etiology of these reactions remained unknown for many years, the causative agent was thought to be endotoxin and confirmation of this hypothesis awaited the advent of the LAL test.[11] Although this assay is sometimes plagued by specificity in special test situations, it is approximately 10 to 100 times more sensitive than the conventional rabbit pyrogen test.[12]

Using the LAL test for endotoxin, Raji et al. demonstrated endotoxin in 41% of those patients having febrile reactions, while 66 samples taken from afebrile patients exhibited no LAL reactivity. These observations clearly implicate endotoxin in febrile reactions of hemodialysis patients. Since patients served as their own controls (febrile and afebrile), these observations are rather compelling in support of exogenous endotoxin being the etiologic agent of these reactions. Similar experiences were reported by Robinson et al.[77]

Additional studies in which the LAL test was used as a measure of endotoxin provide further evidence that the etiologic agent of hemodialysis-related febrile reactions is endotoxin produced from Gram-negative bacteria.[13-15] Additional data based on the ability to terminate outbreaks of pyrogenic reactions by reducing Gram-negative bacterial or endotoxin contamination in fluids associated with dialysis procedures further suggest that under certain conditions endotoxin in the dialysis fluids can pass through the dialysis membrane in the absence of apparent leaks.[15,16,18-20]

C. MEMBRANE PERMEATION OF ENDOTOXIN

Although there are unequivocal associations implicating endotoxin-contaminated water and dialysate with pyrogenic episodes in patients, the permeation of natural endotoxin through hemodialysis membranes remains controversial.[17] This controversy is rooted in the small pore size of most membranes (30 to 60 Å vs. the relatively large size of endotoxin in an aqueous environment which approximates 100 to 1000 Å).[8,9] Although purified endotoxin (lipopolysaccharide [LPS]) exhibits a highly aggregated state in the presence of divalent cations (dialysate), approximating 1 million Da, fragments can exist which are much smaller or much larger, depending on the environment. Calcium salts of purified endotoxin tend to be insoluble while triethylamine derivatives are extremely soluble.[21] Purified endotoxin can be disaggregated to smaller component parts by surfactants such as sodium cholate or other detergents. Moreover, purified *Escherichia coli* endotoxin will pass through a 10^6 nominal molecular weight ultrafilter in the presence of 0.5% cholic acid, but similar preparations made in the presence of monovalent cations (0.9% NaCl) or divalent cations (5 mm $MgCl_2$) will not pass a 10^6 Da ultrafiltration membrane.[22] However, in the same study, disaggregated forms of endotoxin were retained by a 10^4 Da ultrafilter but not by a 10^5, suggesting the subunit of purified endotoxin to be in excess of 10,000 but less than 100,000 Da. It is interesting to note that there is an inverse relationship between the aggregation state and pyrogenicity, i.e., as aggregation state increases, pyrogenicity decreases. Furthermore, rabbit pyrogenicity and LAL activity both increase as aggregation states decrease.[21]

Additional studies by Man and co-workers[8] examined the possibility of "natural" endotoxin crossing from the dialysate side through the membrane to the blood side during the use of high-flux membrane dialyzers. These membranes have a porosity approximately twice that of a conventional cellulose membrane. Based on endotoxin evaluations using the LAL test, an endotoxin mass transfer rate of 1.84 ± 0.05 ng/min was calculated. Extrapolation from clearance data vs. solute molecular weight of urea and vitamin B_{12} for the polyacrylonitrile membrane suggested an apparent molecular weight of 12,000 Da for endotoxin under these conditions.

These experimental findings are consistent with the occurrence of pyrogenic reactions in patients. The threshold pyrogenic dose for *E. coli* endotoxin is widely recognized as approximately 1.0 ng/kg when administered as a bolus dose.[17] Since 1 to 2 ng/min of endotoxin was transferred under the experimental conditions employed, a total dose of 180 to 480 ng of endotoxin would have been administered (2.6 to 6.9 ng/kg given a 70-kg individual). Transfer of this dose of endotoxin from the dialysate side to the blood path could account for pyrogenic reactions in patients.

Nonetheless, the controversy regarding the permeation of endotoxin across various dialyzer membranes continues. Additional studies using polysulfone membranes failed to show demonstrable crossover of endotoxin in the systems utilized.[23,24] Comparable results have been reported using a variety of cellulose membranes.[25]

D. GENERATION OF INTERLEUKIN-1 BY ENDOTOXIN IN HEMODIALYSIS

Overt pyrogenic reactions represent only one aspect of the potential complications for patients on maintenance dialysis. Endotoxin or its purified counterpart, LPS, is a classic inducer of interleukin-1 (IL-1). The host responds to inflammatory mediators with a remarkably consistent set of changes which are commonly grouped together as the acute phase response. Many of the clinical and laboratory changes noted in chronic hemodialysis are indicative of an acute phase response and are probably mediated through IL-1. Some of these biological activities include pyrogenicity, changes in plasma divalent cations, synthesis of acute phase proteins, skeletal proteolysis, and induction of sleep.[26] These activities are consistent with endotoxin playing a central role in hemodialysis-related dysfunctions.

E. INTERLEUKIN-1

One of the many biological activities attributed to endotoxin is the generation of IL-1, a 15,000 to 70,000 Da human protein which also has the capacity to induce a number of physiological perturbations, including fever (Figure 1). The multiple biological effects of IL-1 have been captured by investigators who assigned various names to these proteins prior to the official recognition that IL-1 was responsible for numerous activities. Depending on the biological assay used to measure IL-1 activity, various names emerged, including endogenous pyrogen (EP), lymphocyte activating factor (LAF), B-cell activating factor (BAF), and leukocytic endogenous mediator (LEM).[27]

A possible relationship between endotoxin and IL-1 in hemodialysis was first advanced by Henderson and co-workers[28] as "The Interleukin Hypothesis". Simply stated, this hypothesis predicts that monocytes capable of producing IL-1 can be absorbed to the blood side of the hemodialysis membrane and stimulated by endotoxin from the dialysate side of the membrane. The net clinical result would be fever, frequently coupled with hypotension. During the past 5 years, a significant amount of work has been conducted to evaluate this hypothesis. On balance, most studies have failed to show movement of endotoxin through the membrane from the dialysate side to the blood side. Nonetheless, *in vitro* studies have demonstrated the generation of IL-1 when endotoxin is placed on the dialysate side of the membrane.[26,30] The mechanism by which this interaction between endotoxin and monocyte occurs is still subject to speculation,

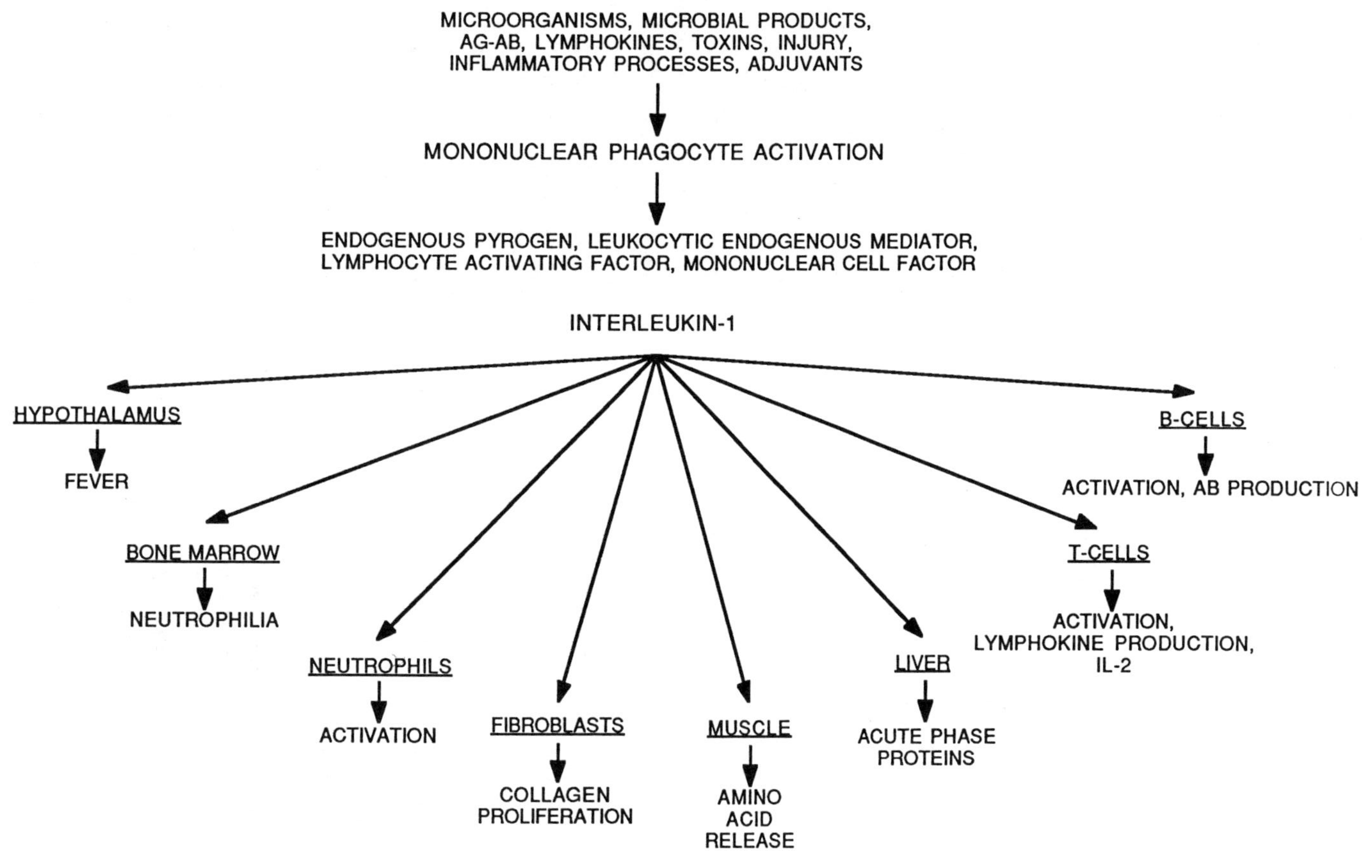

FIGURE 1. The multiple biologic activities of IL-1. AG denotes antigen, AB antibody, and IL-2 (Interleukin-2). (From Dinarello, C., *N. Engl. J. Med.*, 311(22), 1413, 1984. Reprinted with permission from The New England Journal of Medicine.)

since a typical regenerated cellulose membrane separates endotoxin from monocyte by 15 μm.[29] One obvious explanation could be microruptures in the membrane, which would permit the passage of endotoxin while preventing the passage of erythrocytes to the dialysate side. Proctor and co-workers have provided an alternative explanation, since they clearly demonstrated endotoxin subunits of 1000 to 10,000 Da capable of inducing clinical endotoxemia and thus febrile reactions.[31] The pathogenesis of fever in hemodialysis has been reviewed recently.[32]

F. PYROGEN EVALUATION OF WATER IN HEMODIALYSIS

During the past decade, the LAL test has, by and large, replaced the U.S. Pharmacopeia (USP) rabbit pyrogen test for the evaluation of water, medical devices, intravenous solutions, parenterals, radiopharmaceuticals, and some biologicals.[33] The LAL test has revolutionized pyrogen testing worldwide, due to the pioneering observations of Levin and Bang[34] and the subsequent commercial development of specific and sensitive LAL reagents. The LAL test has provided a simple, cost-effective, reproducible endotoxin test that can accurately measure subpyrogenic levels of endotoxin.

Because of its simplicity and sensitivity, the LAL test for endotoxin is ideally suited for evaluation of water prepared for use in hemodialysis. Technically, pure water is perhaps the most easily tested solution since it does not cause LAL test inhibition or enhancement and requires no alteration prior to testing. Moreover, when the first attempts were made to correlate and validate the LAL test to the standard rabbit pyrogen test, water analyses were used as a model system.[35] In this study, the results of over 500,000 LAL analyses were reported and in no case was a rabbit pyrogen failure reported that was not predicted by the LAL test. In fact, these data suggest that with the LAL reagent used, the *in vitro* test overpredicted pyrogenicity, thereby providing a built-in "safety factor". Subsequent studies further suggest that there is a significant difference between "environmental" or real-world endotoxins and commercial preparations of endotoxin with regard to pyrogenicity and LAL reactivity. When commercial endotoxin was spiked in water and administered at an approximate dose of 1 ng/kg, typically the minimal pyrogenic dose (MPD) was reached. In contrast, when tap water samples containing naturally occurring endotoxins as defined by the LAL test were evaluated, the MPD was between 8 to 16 ng/kg. These findings suggest that with the LAL preparation used, the LAL test overpredicted pyrogenicity by a factor of ten. The concept of naturally occurring endotoxin being less pyrogenic than predicted by the LAL test has been confirmed and extended through the use of tap water, ocean water, plasma fractions, and selected drug products.[36]

On balance, these studies suggest that when conducting experimental studies to simulate real world conditions, consideration should be given to using a realistic source of "environmental" endotoxin in the experimental design. Clearly, purified endotoxin (LPS) and environmental endotoxin differ in major ways including purity, composition, and biological activity.

G. WATER QUALITY STANDARDS FOR HEMODIALYSIS REUSE

Hemodialyzers are manufactured and sold as single-use, nonpyrogenic, sterile, and disposable medical devices. Due to the expense of these devices, third-party reimbursement issues, biocompatibility concerns, and first-use syndrome (anaphylaxis), hemodialyzers are frequently reprocessed and commonly reused as many as a dozen times on the same patient. In recent years, the reported incidence of hemodialyzer reuse has risen dramatically, from an estimated 16% of the end-stage renal disease patients to approximately 70% of the patients in 1985.[37] With this practice, concern has developed regarding the proper procedures for reprocessing hemodialyzers, with special emphasis on pyrogenicity and disinfection.

In response to these concerns, the Association for the Advancement of Medical Instrumentation (AAMI) addressed the need for recommended practice for reuse of hemodialyzers.[37] These recommendations have become more commonly known as the AAMI Guidelines for Reuse. Guidelines for water quality are a major focus of the document, which addresses issues such as the water generating system, reprocessing system, utility requirements, validation testing, maintenance, and environmental control.

To be in compliance with the guidelines, tap water used to rinse and clean a used dialyzer should be filtered through a nominal 5 μm filter and have a bacterial colony count of less than 200/ml and/or an endotoxin concentration of less than 1 ng/ml as measured by the LAL test. These same microbial and endotoxin limit standards apply to the final solution. In addition, the diluent (water) used to prepare the germicide for disinfection of the used dialyzer should have a bacterial colony count of less than 200/ml and/or an endotoxin content of less than 1 ng/ml. Using the newer designation, this figure translates to approximately 5 endotoxin units (EU)/ml, depending on the endotoxin standard used to evaluate the potency of the particular LAL reagent. Typically, the manufacturer of the LAL reagent will list this information on the package insert.

The guidelines further provide for the monitoring of patients during dialysis, with special emphasis on fever and chills. The guideline calls for measuring and recording temperature at least prior to and after dialysis with both new and reused dialyzers. A temperature of over 37.8°C (100°F) taken orally should be reported to a physician. When unexplained fever or chills occur more frequently with used dialyzers than with new dialyzers, this observation should serve as a signal to reevaluate the water being used for reprocessing with the LAL test to ensure the proper water quality.

It is clear that high levels of endotoxin can be retained by hemodialyzers during reprocessing, and the subsequent elevation of endotoxin by blood during hemodialysis can lead to profound endotoxic reactions. To prevent these reactions, a system must be used to provide adequate water quality along with frequent monitoring of water with the LAL test.

H. DIALYSATE

As described previously, dialysate is a hyperosmolar solution containing various inorganic ions and small molecular weight, organic, carbon-containing compounds such as acetate or bicarbonate. While high concentrations of these substances aid in purification of uremic blood, these preparations can be a reservoir for clinically significant amounts of endotoxin. Despite the controversy surrounding the passage of endotoxin from the dialysate side to the blood side of the hemodialyzer, both laboratory studies and clinical observations underscore the danger of having excessively high levels of endotoxin in dialysate.

The introduction of an acetate-based dialysate was made in 1964 to obviate the instability associated with bicarbonate dialysate.[38] Subsequently, acetate-based dialysate has been implicated in several dialysis-associated patient complications, including hemodynamic instability, hyperlipidemia, and buffer acidification.[39-42] Since numerous clinical studies have demonstrated the superior biocompatibility of bicarbonate compared to acetate dialysate, initiatives have been undertaken to develop a stable bicarbonate concentrate and techniques for its storage and delivery. These efforts have been successful and bicarbonate dialysate is widely used today.

Unfortunately, bicarbonate dialysate has been associated clinically with multiple outbreaks of overt pyrogenic reactions.[8,43,44] Unlike acetate concentrate, liquid bicarbonate supports the growth of Gram-negative endotoxin containing bacteria, particularly *Pseudomonas*.[45] In part, the explanation for this is that the molarity of the bicarbonate is $^1/_4$ that of acetate concentrate (1.2 vs. 4.8*M*). Using Standard Methods Agar, supplemented with 0.2 *M* NaCl and 0.4 *M* $NaHCO_3$, viable organisms were consistently isolated from bicarbonate concentrate (range 100 to 320,000/ml). In the same study, endotoxin levels were determined to range from nondetectable (<0.2 ng/ml) to approximately 200 ng/ml. When clear liquid bicarbonate was used as a growth medium for Gram-negative bacterial isolates, a 3 to 5 log increase was seen at 72 h along with endotoxin levels as high as 50 ng/ml. Studies by others have confirmed and extended these findings. Liquid bicarbonate supports not only microbial growth, but also the generation of endotoxin.[46]

Since bacteremia and pyrogenic reactions have been associated with the use of water and

TABLE 1
Endotoxin Levels in Diasol® as Measured by the Chromogenic LAL Assay (FDA limit = none; n = 16)

	Range	Average
Endotoxin level, EU/ml	0.01—0.16	0.05
Endotoxin level, ng/ml	0.002—0.03	0.01

From Pearson, F. C., Dubczak, J., Weary, M., and Anderson, J., *Blood Purif.*, 6, 207, 1988. With permission from S. Karger AG, Basel, Switzerland.

dialysate that exceeded the AAMI bacteriologic standard, AAMI recommends that dialysate be cultured at least monthly to ensure that levels are not higher than their maximum recommended level of 2000 colony-forming units (CFU)/ml.[43,44]

Man and associates[8] have reported that 300 microorganisms are commonly found in 100 g of sodium bicarbonate salt, whereas between 10 and 40 microorganisms are demonstrable in 100 g of potassium chloride and sodium chloride, respectively. With this high level of bioburden in the raw material, one would expect both a high bioburden as well as a high pyroburden in the final use product. Bacteriological examination of the bicarbonate-containing concentrate revealed 10^{10} to 10^{11} *Pseudomonas aeruginosa* per milliliter of concentrate. Prospective use of this material clinically resulted in a series of adverse reactions when patients were dialyzed on high-flux polyacrylonitrile membrane dialyzers. Adverse reactions were not associated with bicarbonate dialysate and Cuprophane membrane dialyzers. Since this observation, the French GMP has mandated that neither bicarbonate nor acetate concentrates can contain more than 10 viable counts per milliliter of halophilic species of bacteria and less than 500 viable counts per milliliter for mesophilic species.[8]

Since dialysis with bicarbonate containing dialysate is more widely used in high-risk patients with poor hemodynamic conditions, monitoring of final dialysate is warranted, particularly with high-flux membranes. While microbial standards have been set for final-use dialysate, abrupt changes in microbial levels should be a signal for concern. Moreover, routine monitoring of dialysate for fluctuations in endotoxin with LAL could provide a more accurate, simple, and rapid method for predicting impending adverse clinical reactions. Dialysate is readily examined for endotoxin by the LAL test. Care must be taken to dilute dialysate with nonpyrogenic water to overcome any test inhibition (high molarity) or enhancement due to divalent cations in the dialysate. Very low levels of endotoxin have been reported for properly manufactured dialysate (see Table 1).[9,47]

I. ENDOTOXIN EVALUATION OF HEMODIALYZER

Hemodialyzers are marketed as sterile, nonpyrogenic, and disposable medical devices. Commercial pyrogen evaluation of hemodialyzers is conducted by either the rabbit pyrogen test or the LAL test. Typically each AK to be tested is filled with nonpyrogenic water or saline (rabbit) and the eluates from 10 units are collected, pooled, and examined for endotoxin. Dialyzers prepared under GMP have very low subpyrogenic levels of endotoxin associated with the blood path. Since the test solution for endotoxin is a water or saline rinse of the blood path, the LAL test is relatively straightforward and does not suffer from test inhibition or enhancement, with one major exception.

In 1982, two separate groups observed high levels of "endotoxin" associated with some hemodialyzer membranes.[48,49] Although pyrogenic levels of "endotoxin" were demonstrated in these dialyzers, there was no demonstrable pyrogenicity in rabbits or humans that could be associated with the high LAL results. Thus, this LAL-reactive material became known as LAL-RM. In a subsequent series of studies, LAL-RM was shown to be a nonspecific activator of some

TABLE 2
LAL-RM Associated with Various Cellulose-Based Hollow-Fiber Dialyzers

	LAL-RM[a]	Sensitivity[b]
Nipro cellulose acetate FB90	12.6	<6.25
C-Dak 400 cellulose acetate	12.3	<6.25
C-Dak saponified cellulose acetate	5960	<6.25
Travenol 1787	5366	<6.25

[a] Referenced to picogram per milliliter of *E. coli* 055:B5 endotoxin.
[b] The label sensitivity of the preparation was 6.25 pg/ml.

From Pearson, F. C., *Blood Purif.*, 5, 115, 1987. With permission from S. Karger AG, Basel, Switzerland.

commercial LAL reagents and associated with only some kinds of membrane materials.[50,56] These relationships can be seen in Table 2. LAL-RM is associated with Cuprophane, a regenerated cellulose membrane, and saponified cellulose acetate, but not with cellulose acetate. LAL-RM has not been demonstrably associated with polymethylmethacrylote, ethyl vinyl alcohol, polyacrylonitrile, or polysulfone membranes.[76] One can also see in Table 2 that one commercial LAL reagent did not detect LAL-RM, while another reagent was sensitive to this nonspecific activating material. The table also shows that exceedingly low levels of endotoxin are associated with the membranes evaluated.

Through the use of size-exclusion chromatography and gas chromatography/mass spectrometry, LAL-RM was found to be a cellulose-based material having an average molecular weight of 24,000 Da.[50] Unlike endotoxin, this material did not induce IL-1 generation, cause lymphocyte or macrophage activation, or induce the release of arginase from macrophages. LAL-RM failed to induce leukopenia followed by rebound leukocytosis or alterations in coagulation and blood chemistries as would be expected for endotoxin.

In a series of elegant studies conducted by Yamagami and associates, two clotting cascades were elucidated in the LAL reagent. The conventional enzymatic cascade is initiated by the interaction of endotoxin with an enzyme called factor C. Activated factor C then activates factor B which turns on the downstream enzymatic reaction leading to the LAL endpoint. An alternate cascade that also turns on the downstream cascade is initiated by the interaction of factor G and (1–3)-β-D-glucan (LAL-RM). These cascades can be seen in Figure 2 (see also Chapter 3).

Since LAL-RM is membrane-associated and elutes over time, LAL testing of reprocessed dialyzers could reveal falsely high levels of endotoxin. When a new hemodialyzer is filled with nonpyrogenic water and is allowed to stand overnight, appreciable amounts of LAL-RM will be extracted. If the dialyzer is rinsed rapidly until free of LAL-RM, refilled with nonpyrogenic water, and again allowed to stand overnight, the LAL-RM level will rebound. This procedure can be repeated for 7 d, and significant elution of LAL-RM is still apparent. Studies show that patients undergoing dialysis have large amounts of plasma LAL-RM predialysis, which escalates by 50% after 15 min of dialysis.[53] Thus, blood would appear to be a good extractant for LAL-RM. The clinical significance of this observation is still open to question.

When evaluating Cuprophane or saponified cellulose acetate membranes for endotoxin, it is critical to use a LAL reagent that does not react with LAL-RM unless the blood path has been cleansed of this material. Currently, there are several manufacturers of LAL reagents that allegedly do not give false positives with hemodialyzer membranes. Of particular note is the Seikagaku Kogyo Co., in Japan. Their preparation has had the factor G cascade removed, thereby obviating false-positive reactions. This test is in a chromogenic format and, in the author's experience, has given results consistent with the claims of the manufacturer.

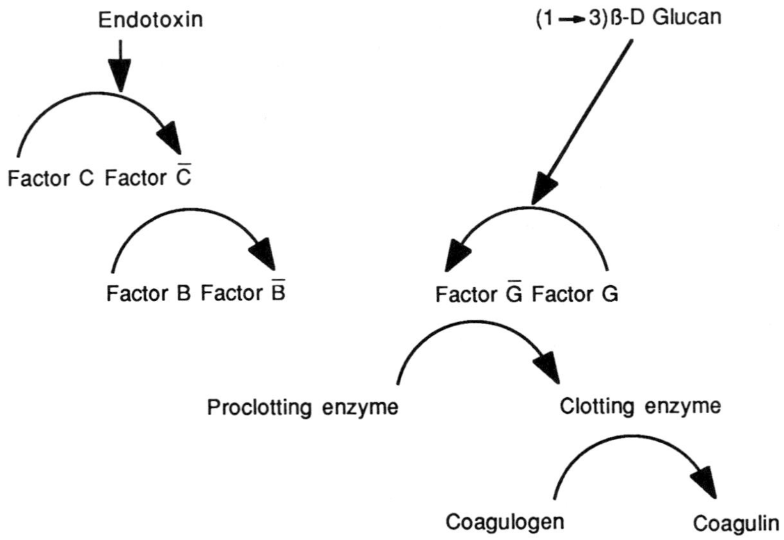

FIGURE 2. Cascade reaction of the *Limulus* test. (From Yamagami, S., Yoshimura, R., Sugimura, K., Yoshiwara, H, Kishimoto, T., Maekawa, M., and Niwa, M., *Jpn. J. Artif. Organs*, 15(3), 1327, 1986. With permission.)

III. SURVEILLANCE OF RESPIRATORY EQUIPMENT

While the detection of endotoxin associated with hemodialysis systems represents a major application for the LAL test, other allied medical applications include endotoxin monitoring of humidifier systems, food, and water. Each of these areas are discussed here.

High levels of airborne endotoxin have been demonstrated in a number of industrial environments, particularly where cotton and flax are processed.[57] As many as 10^2 to 10^9 Gram-negative bacteria (the only source of endotoxin) have been demonstrated per gram of material. During normal working conditions, as much as 24 μg/M^3 of endotoxin has been detected in the air. The clinical manifestations of this respiratory syndrome became so commonplace in textile workers that it became known as byssinosis, derived from the Greek word "byssor" meaning cotton.

Since endotoxin is highly dispersed in air, and since pulmonary beds provide a very large surface area, endotoxin is delivered through the lungs efficiently, culminating in local as well as systemic pathophysiology. Exposure to airborne endotoxin causes pulmonary inflammation, as evidenced by infiltration of polymorphonuclear leukocytes (PMN) in both humans and animals. These observations have been made in a number of studies conducted by Rylander and others.[57,59] Evidence for the systemic impact of endotoxin delivered through the lung is supported by the observation of fever, which is generated most commonly through blood-borne monocytes and macrophages. Additional studies in animals subjected to a 1-ml aerosolized endotoxin challenge of 1 μg/ml caused decreased lung diffusion capacity for up to 6 h.[58]

Clearly, both the laboratory and clinical experimental data underscore the potential danger of delivering unwarranted levels of endotoxin through the inadvertent use of respiratory equipment such as nebulizers and humidifiers.

Rylander and co-workers[60] reported three cases of "humidifier fever" associated with Flavobacterium-contaminated humidifier water. Subsequently, clinically reactive individuals were exposed to effluent from the humidifier and exhibited responses that were consistent with exposure to endotoxin. These symptoms included fever, respiratory dysfunction, rebound leukocytosis, and an increase in the number of segmented white blood cells (WBC). From a pragmatic standpoint, these observations suggest that humidifiers must be cleaned on a regular basis to circumvent microbial growth and the concomitant generation of endotoxin from Gram-negative bacteria. Companion papers have reported *Cytophoza* sp. endotoxin as a putative agent of occupational lung disease.[61,62]

The significance of these observations was confirmed and extended by Reinhardt and associates.[63] In an evaluation examining over 200 items of respiratory therapy, the most frequently contaminated devices were medication cups of ultrasonic nebulizers and reservoirs from bubblers and nebulizers. Quantitatively, contamination of these devices and solutions ranged from 10 to 500,000 CFU/ml. Qualitatively, contaminants were comprised of typical Gram-negative water bacteria including *Pseudomonas* and *Acinetobacter* sp. Identification of both numbers and kinds of bacterial contaminants present provides the opportunity to eliminate or reduce bacterial endotoxin through appropriate disinfection and sterilization. Since speciation of bacteria and their enumeration is labor-intensive and requires 24 to 48 h, the LAL test became an ideal tool to monitor respiratory equipment for Gram-negative bacterial contamination (endotoxin).

Parallel evaluations comparing direct sampling methods and the LAL test for surveillance of operating nebulizers were undertaken in a subsequent study by Reinhardt and co-workers.[64] The LAL test was shown to be sensitive to approximately 10^3 CFU/ml, which represents a moderate to heavy microbial contamination of the in-use solutions examined. While the Easicult (Orion Diagnostics) system for microbial contamination was sensitive to about 10^3 CFU/ml as well, the LAL test requires only 1 h to complete, and it provides meaningful results. Positive LAL assays coupled with negative culture tests were observed in 20 of 206 samples (9.7%), indicating the presence of endotoxin in the absence of viable bacteria. Conversely, positive culture results and negative LAL results were seen in 6.8% of the sample. These observations demonstrate the potential utility of the LAL test and Easicult for the surveillance of operating respiratory therapy equipment.

Although recommendations have been made to change respiratory equipment every 12 to 24 h, unfortunately some hospitals do not replace nebulizers, tubing, and related equipment this frequently.[65,66] Moreover, abuse of respiratory equipment over this period can occur, thereby leading to increased microbial growth. Proper operation and disinfection, and appropriate surveillance of respiratory equipment cannot be overestimated. A definitive relationship between the incidence of Gram-negative necrotizing pneumonia and the incidence of contaminated nebulizing devices has been reported.[67,68] In fact, hospital-acquired infections of the respiratory tract rank third in the U.S., while urinary tract and surgical wound infections rank higher.[69]

Without question, proper care and use of respiratory equipment is imperative if hospital-acquired respiratory tract infections are to be significantly reduced. The LAL test clearly provides an additional tool for the monitoring of potentially contaminated equipment and solutions. As stated before, the LAL test is sensitive, specific, simple, and rapid, and is a relatively inexpensive assay to perform compared to bacterial culture.

IV. DETECTION OF CONTAMINATED FOODS AND WATER

Since the microbial spoilage of refrigerated ground beef is caused primarily by Gram-negative bacteria, the LAL test has provided some advantages not available through other tests.[70] Jay reported that he found endotoxin concentrations to increase from 10^2 to 10^5 in refrigerated ground beef over a 7 to 12 d period. These findings correlate significantly with total viable Gram-negative bacterial counts at both 5 and 30°C. LAL assays were negative for fresh beef and aged ground beef containing low numbers of bacteria. In a subsequent study by Fallowfield and Patterson,[71] the findings of Jay were confirmed and extended. With the use of a modified LAL microtiter procedure and a more sensitive commercial LAL reagent, *Pseudomonas* sp. could be detected at 10^2 to 10^3 CFU/ml. Using this procedure, endotoxin levels and microbial cell counts were conducted on ground beef and minced pork. Not only did endotoxin levels increase with the onset of meat spoilage, but factors were derived which provided an estimate of approximate microbial counts (AMC) from the endotoxin concentrations.

When the LAL test was compared to chemical analyses for the assessment of the quality of lean fish, similar results were observed.[72] Endotoxin levels agreed with aerobic plate counts and chemical indices of fish spoilage. The correlation between the level of endotoxin and the level of total volatile bases was highly significant ($r = 0.8574$; $p \leq 0.001$). The LAL test has been used also for the detection of endotoxin in water, including waste water, fresh water, and marine water.[73-75]

The LAL test has also been evaluated for use in detecting mastitis in cattle and contaminated processed milk and milk products. See Chapter 14 for a detailed discussion of this topic.

V. CONCLUSIONS

Over the last 2 decades, the original observations of Levin and Bang have developed into the most sensitive, specific, simple, rapid, accurate, and inexpensive assay for endotoxin currently available. Not only has the LAL test largely replaced the more cumbersome rabbit pyrogen test, but it has made monitoring for endotoxin contamination practical in a number of situations not heretofore possible. Clearly, the LAL test has made the surveillance of endotoxin practical in hemodialysis centers, and in the evaluation of hemodialyzer reuse, nebulizers, water, and food.

REFERENCES

1. **Pollak, V. E.,** Adverse effects and pyrogenic reactions during hemodialysis, *JAMA*, 260(14), 2107, 1988.
2. **Gordon, S. M., Tipple, M., Bland, L. A., and Jarvis, W. R.,** Pyrogenic reactions associated with the reuse of disposable hollow-fiber hemodialyzers, *JAMA*, 260(14), 2077, 1988.
3. **Alter, M. J., Favero, M. S., Miller, J. K., Coleman, P. J., and Bland, L. A.,** Reuse of hemodialyzers. Results of nationwide surveillance for adverse effects, *JAMA*, 260(14), 2073, 1988.
4. **Pearson, F., Bruszer, G., Lee, W., Sagona, M., Sargent, H., Woods, E., Dolovich, J., and Caruana, R.,** Ethylene oxide sensitivity in hemodialysis patients, *Artif. Organs*, 11(2), 100, 1987.
5. **Chenoweth, D. E.,** Complement activation during hemodialysis: clinical observations, proposed mechanisms, and theoretical implications, *Artif. Organs*, 8(3), 281, 1984.
6. **Bok, D. V., Pascual, L., Herberger, C., Sawyer, R., and Levin, N. W.,** Effect of multiple uses of dialyzers on intradialytic symptoms, *Proc. Clin. Dial. Transplant Forum*, 10, 92, 1980.
7. **Kant, K. S., Pollok, V. E., Cathey, M., Goetz, D., and Berlin, R., with technical assistance of Eydell, D. F., Tessel, J., and Hall, R.,** Multiple use of dialyzers: safety and efficacy, *Kidney Int.*, 19, 728, 1981.
8. **Man, N. K., Ciancioni, C. H., Guyomard, S., Blais, D., Delons, S., and Funck-Brentano, J. L.,** Risks and hazards of contaminated dialysate associated with high flux membranes, in *Prevention in Nephrology*, Biccoamto, G., Ed., Masson Italia Editori, Milano, 1987, 227.
9. **Pearson, F. C., Dubczak, J., Weary, M., and Anderson, J.,** Determination of endotoxin levels and their impact on interleukin-1 generation in continuous ambulatory peritoneal dialysis and hemodialysis, *Blood Purif.*, 6, 207, 1988.
10. **Raij, L., Shapiro, F. L., and Michael, A. F.,** Endotoxemia in febrile reactions during hemodialysis, *Kidney Int.*, 4, 57, 1973.
11. **Levin, J. and Bang, F. B.,** Clottable protein in Limulus: its localization and kinetics of its coagulation by endotoxin, *Thromb. Diath. Haemorrh.*, 19, 186, 1968.
12. **Pearson, F.,** A comparison of the pyrogenicity of environmental endotoxins and lipopolysaccharides; in *Bacterial Endotoxins: Structure, Biomedical Significance and Detection with the Limulus Amebocyte Lysate Test*, Ten Cate, J. W., Büller, H. R., Sturk, A., and Levin, J., Eds., Alan R. Liss, New York, 1985, 251.
13. **Carson, L. A., Petersen, N. J., and Favero, M. S.,** Use of the Limulus amoebocyte lysate assay system for detection of bacterial endotoxin in fluids associated with hemodialysis procedures, in *Biomedical Applications of the Horseshoe Crab (Limulidae)*, Cohen, E., Ed., Alan R. Liss, New York, 1979, 453.

14. **Favero, M. S., Carson, L. A., Bond, W. W., and Petersen, N. J.,** Factors that influence microbial contamination of fluids associated with hemodialysis machines, *Appl. Microbiol.*, 28, 822, 1974.
15. **Petersen, N. J., Boyer, K. M., Carson, L. A., and Favero, M. S.,** Pyrogenic reactions from inadequate disinfection of a dialysis fluid distribution system, *Dial. Transplant.*, 7, 52, 1978.
16. **Carson, L. A. and Petersen, N. J.,** LAL-reactive material associated with hemodialysis membranes, in *Endotoxins and Their Detection with the Limulus Amebocyte Lysate Test*, Watson, S. W., Levin, J., and Novitsky, T. J., Eds., Alan R. Liss, New York, 1982, 217.
17. **Greisman, S. E. and Hornick, R. B.,** Comparative pyrogenic reactivity of the rabbit and man to bacterial endotoxin, *Proc. Soc. Exp. Biol. Med.*, 131, 1154, 1969.
18. **Petersen, N. J., Carson, L. A., Favero, M. S., Marshall, J. H., and Aguero, S. M.,** Removal of bacteria and bacterial endotoxin from dialysis fluids by sorbents, *Am. Soc. Artif. Intern. Organs*, 3, 6, 1980.
19. **Favero, M. S.,** *Microbiological Contaminants: Issues in Hemodialysis*, Association for the Advancement of Medical Instrumentation, Arlington, VA, 1981.
20. **Favero, M. S.,** Distinguishing between high-level disinfection, reprocessing, and sterilization, in *Reuse of Disposables: Implications for Quality Health Care and Cost Containment*, Association for the Advancement of Medical Instrumentation, Arlington, VA, 1983, 19.
21. **Sweadner, K.J., Forte, M., and Nelsen, L.,** Filtration removal of endotoxin (pyrogens) in solution in different stages of aggregation, *Appl. Environ. Microbiol.*, 34, 382, 1977.
22. **Galanos, C., Freudenberg, M. A., Luderitz, O., Rietschel, E. T., and Westphal, O.,** Chemical, physiochemical and biological properties of bacterial lipopolysaccharide, in *Biomedical Applications of the Horseshoe Crab (Limulidae)*, Cohen, E., Ed., Alan R. Liss, New York, 1979, 319.
23. **Bommer, J., Becker, K. P., Urbaschek, R., Ritz, E., and Urbaschek, B.,** No evidence for endotoxin transfer across high flux polysulfone membranes, *Clin. Nephrol.*, 27, 278, 1987.
24. **Klinkmann, H., Falkenhagen, D., and Smollich, B. P.,** Investigation of the permeability of highly permeable polysulfone membranes for pyrogens, *Contrib. Nephrol.*, 46, 174, 1985.
25. **Port, F. K. and Bernick, J. J.,** Pyrogen and endotoxin reactions during hemodialysis, *Contrib. Nephrol.*, 36, 100, 1983.
26. **Dinarello, C. A.,** The biology of Interleukin-1 and its relevance to hemodialysis, *Blood Purif.*, 1, 197, 1983.
27. **Krakauer, T.,** Human Interleukin 1, *Crit. Rev. Immunol.*, 6(3), 213, 1986.
28. **Henderson, L. W., Koch, K. M., Dinarello, C. A., and Shaldon, S.,** Hemodialysis hypotension: the Interleukin hypothesis, *Blood Purif.*, 1, 3, 1983.
29. **Bingel, M., Lonnemann, G., Shaldon, S., Koch, K. M., and Dinarello, C. A.,** Human Interleukin-1 production during hemodialysis, *Nephron*, 43, 161, 1986.
30. **Shaldon, S., Deschodt, G., Branger, B., Granolleras, C., Baldamus, C. A., Koch, K. M., Lysaght, M. J., and Dinarello, C. A.,** Haemodialysis hypotension: the Interleukin hypothesis restated, unpublished results.
31. **Proctor, R. A., Burhop, K. E., Helgerson, R., Raetz, C. R. H., and Will, J. A.,** Lipid HX: a monosaccharide subunit of endotoxin capable of producing endotoxemia, *Clin. Res.*, 31, 498, 1983.
32. **Dinarello, C. A.,** Pathogenesis of fever during hemodialysis, *Contrib. Nephrol.*, 36, 90, 1983.
33. **Pearson, F. C., Weary, M. E., and Dabbah, R.,** A corporate approach to in-process and end-product testing with the LAL assay for endotoxin, in *Endotoxins and Their Detection with the Limulus Amebocyte Lysate Test*, Watson, S. W., Levin, J., and Novitsky, T. J., Eds., Alan R. Liss, New York, 1982, 231.
34. **Levin, J. and Bang, F.,** The role of endotoxin in the extracellular coagulation of Limulus blood, *Bull. Johns Hopkins Hosp.*, 115, 337, 1964.
35. **Pearson, F. C., Weary, M. E., Bohon, J., and Dabbah, R.,** Relative potency of "Environmental" endotoxin as measured by the Limulus amebocyte lysate test and the USP rabbit pyrogen test, in *Endotoxins and Their Detection with the Limulus Amebocyte Lysate Test*, Watson, S. W., Levin, J., and Novitsky, T. J., Eds., Alan R. Liss, New York, 1982, 65.
36. **Pearson, F. C.,** A comparison of the pyrogenicity of environmental endotoxins and lipopolysaccharides, in *Bacterial Endotoxins: Structure, Biomedical Significance, and Detection with the Limulus Amebocyte Lysate Test*, Ten Cate, J. W., Buller, H. R., Stark, A., and Levin, J., Eds., Alan R. Liss, New York, 1985, 251.
37. *Reuse of Hemodialyzers*, Association for the Advancement of Medical Instrumentation,Arlington,VA,1986, 1.
38. **Mion, C. M., Hegstrom, R. M., Boen, S. T., and Scribner, B. H.,** Substitution of sodium acetate for bicarbonate in the bath fluid for hemodialysis, *Trans. Am. Soc. Artif. Intern. Organs*, 10, 110, 1964.
39. **Keshaviah, P., Ed.,** Investigation of the Risks and Hazards Associated with Hemodialysis Devices, Food and Drug Administration Publication No. 0-625-146/1864, U.S. Government Printing Office, Silver Springs, MD, 1980, 75.
40. **Novello, A., Kelsch, R. C., and Esterling, R. E.,** Acetate intolerance during hemodialysis, *Clin. Nephrol.*, 5, 29, 1976.
41. **Iseki, K., Onoyama, K., Maeda, T., Shimamatsu, K., Harada, A., Fujimi, S., and Omae, T.,** Comparison of hemodynamics induced by conventional acetate hemodialysis, bicarbonate hemodialysis and ultrafiltration, *Clin. Nephrol.*, 14, 294, 1980.

42. **Tolchin, N., Roberts, J. L., Hayaski, J., and Lewis, E. J.,** Metabolic consequences of high mass transfer hemodialysis, *Kidney Int.*, 11, 366, 1977.
43. **Favero, M. S., Petersen, N. J., Carson, L. A., Bond, W. W., and Hindman, S. H.,** Gram negative water bacteria in hemodialysis systems, *Health Lab. Sci.*, 12, 321, 1975.
44. **Favero, M. S., Petersen, N. J., Boyer, K. M., Carson, L. A., and Bond, W. W.,** Microbial contamination of renal dialysis systems and associated health risks, *Trans. Am. Soc. Artif. Intern. Organs*, 20, 175, 1974.
45. **Bland, L. A., Ridgeway, M. R., Aguero, S. M., Carson, L. A., and Favero, M. S.,** Potential bacteriologic and endotoxin hazards associated with liquid bicarbonate concentrate, *Trans. Am. Soc. Artif. Intern. Organs*, 33, 542, 1987.
46. **Ebben, J. P., Hirsch, D. N., Luehmann, D. A., Collins, A. J., and Keshaviah, P. R.,** Microbiologic contamination of liquid bicarbonate concentrate for hemodialysis, *Trans. Am. Soc. Artif. Intern. Organs*, 33, 269, 1987.
47. **Tominaga, H., Tanaka, S., and Tominaga, N.,** Endotoxin level of sterile injection solutions and substitution fluid for hemofiltration in Japan and Australia, *Nephron*, 42, 128, 1986.
48. **Pearson, F. C., Weary, M., and Bohon, J.,** Detection of Limulus amebocyte lysate reactive material in capillary flow hemodialyzers, in *Endotoxins and Their Detection with the Limulus Amebocyte Lysate Test*, Watson, S. W., Levin, J., and Novitsky, T. J., Eds., Alan R. Liss, New York, 1982, 247.
49. **Carson, L. A. and Petersen, N. J.,** LAL-reactive material associated with hemodialysis membranes, in *Endotoxins and Their Detection with the Limulus Amebocyte Lysate Test*, Watson, S. W., Levin, J., and Novitsky, T. J., Eds., Alan R. Liss, New York, 1982, 217.
50. **Pearson, F. C., Bohon, J., Lee, W., Bruser, G., Sagona, M., Jakubowski, G., Dawe, R., Morrison, D., and Dinarello, C.,** Characterization of *Limulus* amoebocyte lysate-reactive material from hollow-fiber dialyzers, *Appl. Environ. Microbiol.*, 48, 1189, 1984.
51. **Pearson, F. C., Caruana, R., Burkart, J., Katz, D. V., Chenoweth, D., Dubczak, J., Bohon, J., and Weary, M.,** The use of the *Limulus* amebocyte lysate assay to monitor hemodialyzer-associated soluble cellulosic material (LAL-reactive material), in *Detection of Bacterial Endotoxins with the Limulus Amebocyte Lysate Test*, Watson, S. W., Levin, J., and Novitsky, T. J., Eds., Alan R. Liss, New York, 1987, 211.
52. **Carson, L. A., Bland, L. A., Cusick, L. B., Collin, S., Favero, M. S., and Bolan, G.,** Factors affecting endotoxin levels in fluids associated with hemodialysis procedures, in *Detection of Bacterial Endotoxins with the Limulus Amebocyte Lysate Test*, Watson, S. W., Levin, J., and Novitsky, T. J., Eds., Alan R. Liss, New York, 1987, 223.
53. **Pearson, F. C.,** The possible role of Limulus-amebocyte-lysate-reactive material in hemodialysis, *Blood Purif.*, 5, 115, 1987.
54. **Pearson, F. C.,** Relative importance of membrane Limulus amebocyte lysate-reactive material and ethylene oxide in hypersensitivity reactions, *Contrib. Nephrol.*, 59, 134, 1987.
55. **Yamagami, S., Yoshimura, R., Sugimura, K., Yoshiwara, H., Kishimoto, T., Maekawa, M., and Niwa, M.,** Limulus-positive substance in cellulose dialyser, *Jpn J. Artif. Organs*, 15(3), 1327, 1986.
56. **Yamagami, S., Yoshihara, H., Iritani, S., Umeda, M., Senju, M., Kisimoto, T., Niwa, M., Iwanaga, S., and Tanaka, S.,** Is an LAL reactive substance contained in the washing solution of unused dialyzer endotoxin?, *Jpn J. Artif. Organs*, 14(1), 30, 1985.
57. **Rylander, R. and Vesterlund, J.,** Airborne endotoxins in various occupational environments, in *Endotoxins and Their Detection with the Limulus Amebocyte Lysate Test*, Watson, S. W., Levin, J., and Novitsky, T. J., Eds., Alan R. Liss, New York, 1982, 399.
58. **Muittari, A., Rylander, R., and Salkinoja-Salonen, M.,** Endotoxin and bath water fever, *Lancet*, 2, 89, 1980.
59. **Muittari, A., Kuusisto, P., Virtomen, P., Sovijarvi, A., Grönroos, P., Harmoinen, A., Antila, P., and Kellomäki, L.,** An epidemic of extrinsic allergic alveolitis caused by tap water, *Clin. Allergy*, 10, 77, 1980.
60. **Rylander, R., Haglind, R., Lundholm, M., Mattsby, I., and Stenqvist, K.,** Humidifier fever and endotoxin exposure, *Clin. Allergy*, 8, 511, 1978.
61. **Flaherty, D. K., Deck, F. H., Cooper, J., Bishop, K., Winzenburger, P. A., Smith, L. R., Bynum, L., and Witmer, W. B.,** Bacterial endotoxin isolated from a water spray air humidification system as a putative agent of occupation-related lung disease, *Infect. Immun.*, 43(1), 206, 1984.
62. **Flaherty, D. K., Deck, F. H., Hood, M. A., Liebert, C., Singleton, F., Winzenburger, P., Bishop, K., Smith, L. R., Bynum, L. M., and Witmer, W. B.,** A Cytophaga species endotoxin as a putative agent of occupation-related lung disease, *Infect. Immun.*, 43(1), 213, 1984.
63. **Reinhardt, D. J., Kennedy, C., and Maleck-Griggs, B.,** Quantitative and qualitative analyses of microbial contamination in solutions and wet, in-use equipment, *Dev. Ind. Microbiol.*, 19, 377, 1978.
64. **Reinhardt, D. J., Nabors, W., Kennedy, C., and Malecka-Griggs, B.,** *Limulus* amoebocyte lysate and direct sampling methods for surveillance of operating nebulizers, *Appl. Environ. Microbiol.*, 42(5), 850, 1981.
65. **Griggs, G. M. and Reinhardt, D. J., Eds.,** *Fundamentals of Nosocomial Infections Associated with Respiratory Therapy*, 2nd ed., Healthscan, Upper Montclair, NJ, 1971.
66. **Mackel, D.,** Sterilization and disinfection of respiratory therapy equipment, in *Association of Practitioners in Infection Control*, Centers for Disease Control and APIC Newsl., 2, 1, 1974.

67. **Pierce, A. K. and Sanford, J. P.,** Bacterial contamination of aerosols, *Arch. Intern. Med.*, 131, 156, 1973.
68. **Edmondson, E. B. and Sanford, J. P.,** Simple method of bacteriologic sampling of nebulization equipment, *Am. Rev. Respir. Dis.*, 94, 450, 1966.
69. **Reinarz, J. A., Pierce, A. K., Mays, B. B., and Sanford, J. P.,** The potential role of inhalation therapy equipment in nosocomial pulmonary infection, *J. Clin. Invest.*, 44, 831, 1965.
70. **Jay, J. M.,** The Limulus lysate endotoxin assay as a test of microbial quality of ground beef, *J. Appl. Bacteriol.*, 43, 99, 1977.
71. **Fallowfield, H. J. and Patterson, J. T.,** Potential value of the Limulus lysate assay for the measurement of meat spoilage, *J. Food Technol.*, 20, 467, 1985.
72. **Sullivan, J. D., Ellis, P. C., Lee, R. G., Combs, W. S., and Watson, S. W.,** Comparison of the Limulus amoebocyte lysate test with plate counts and chemical analyses for assessment of the quality of lean fish, *Appl. Environ. Microbiol.*, 45(2), 720, 1983.
73. **Jorgensen, J. H., Lee, J. C., Alexander, G. A., and Wolf, H. W.,** Comparison of *Limulus* assay, standard plate count, and total coliform count for microbiological assessment of renovated waste water, *Appl. Environ. Microbiol.*, 37, 928, 1979.
74. **Watson, S. W., Novitsky, T. J., Quinby, H. L., and Valois, F. W.,** Determination of the bacterial number and biomass in the marine environment, *Appl. Environ. Microbiol.*, 33, 940, 1977.
75. **Evans, T. M., Schillinger, J. E., and Stuart, D. G.,** Rapid detection of bacterial water quality by using *Limulus* lysate, *Appl. Environ. Microbiol.*, 35, 376, 1978.
76. **Pearson, F. C.,** unpublished data.
77. **Robinson, P. J. A. and Rosen, S. M.,** Paradoxical reactions during hemodialysis, *Br. Med. J.*, 1, 528, 1971.

Chapter 14

VETERINARY APPLICATIONS — AN OVERVIEW

Edward R. Richter and Emil M. Mikolajcik

TABLE OF CONTENTS

I. Introduction ... 156

II. Mastitis ... 156

III. Milk Quality ... 156

IV. Conclusion ... 157

References ... 157

I. INTRODUCTION

Little research has been conducted on the use of the *Limulus* amoebocyte lysate (LAL) test in veterinary science. What has been reported is primarily concerned with mastitic cows and subsequent milk quality. The sensitive detection of Gram-negative endotoxin (lipopolysaccharide [LPS]) by the LAL test offers the veterinarian or dairy person a rapid and reliable method for the diagnoses of Gram-negative mastitis or determination of poor raw milk quality.

II. MASTITIS

The source of infection in a mastitic cow is her environment. Normally, bacteria enter via the teat canal where they multiply to high numbers in the gland cistern and lactiferous ducts.[1] In addition to the general infection brought about by Gram-negative bacterial infection, the elaboration of large amounts of endotoxin due to bacterial cell death and lysis are absorbed from the udder into the general circulation and add to the clinical findings of toxanemia.[2,3] Schalm et al.[4] demonstrated that clinical symptoms of Gram-negative infection could be reproduced with the use of purified endotoxin, suggesting that it is the endotoxin which contributes to the coliform mastitis syndrome. Hartman et al.[1] reported the successful use of the LAL test in detecting Gram-negative mastitis. Raw milk obtained from *Escherichia coli* infected quarters was diluted above $1{:}10^4$. Samples demonstrating positive LAL clots were correctly correlated with the infected quarters. It is important to note that in many cases of coliform mastitis, milk obtained from infected quarters often contains high levels of serum protein. Serum protein has been shown to interfere with the LAL clotting cascade.[5] To overcome this apparent inhibition, the milk may be extracted with equal portions of chloroform, and the top layer diluted with pyrogen-free water.[1] Hakogi et al.[6] also recently reported the successful use of the LAL test in detecting Gram-negative mastitis. These investigators used a chromogenic LAL test (Toxicolor) and pretreated the samples with percholoric acid to remove interfering factors. Although the use of the LAL test cannot diagnose Gram-positive mastitis, it does offer the clinician a valuable aid in the monitoring of coliform mastitis and assessing the prognosis.[6]

III. MILK QUALITY

Psychrotrophic Gram-negative bacteria are the principal contaminants of raw milk. Pre- and postpasteurization contamination of milk by these bacteria, both on the farm and at the dairy, contribute to the short shelf life of the product. Effective pasteurization normally eliminates Gram-negative bacteria from the milk, although heat stable proteases and lipases elaborated by many of the organisms continue to affect the keeping quality of the finished product. These enzymes are produced by members of the genus *Pseudomonas*, a Gram-negative psychrotrophic bacterium reported to be the predominate contaminant in raw and pasteurized milk.[7,8] Other members of the Gram-negative psychrotrophs include *Alcalgenes, Achromobacter, Aeromonas and Flavobacterium*.[9]

Quantification of Gram-negative psychrotrophic bacteria in raw milk serves two purposes. First, high levels of these organisms in raw milk suggest poor milking sanitation. The poorer the quality of raw milk the shorter the shelf life of the finished product. Second, many dairies have begun paying premiums for raw milk with low bacterial levels. Thus, the better the quality of the raw milk the more valuable the commodity and the more revenue for the milk producer. Currently, standard microbiological methodology is employed to determine the bacterial quality of milk. The aerobic plate count (48 h), psychrotrophic plate count (7 d), and modified psychrotrophic plate count (24 h)[10] are techniques used to enumerate Gram-negative bacterial psychrotrophs in most dairies today. Mikolajcik and Brucker[11] reported on the use of the LAL procedure to determine the quality of raw and pasteurized milk. This method was reported to be more sensitive than the Gram-negative plate count procedure. They reported that raw milks having Gram-negative bacterial counts between 110,000 to 1 million/ml contained endotoxin titers below 100 ng/ml, while milks with levels between 1.1 million to 10 million/ml yielded endotoxin titers below 1000 ng/ml and milks with bacterial levels between 11 million to 100 million/ml correlated with endotoxin titers of less than 10,000 ng/ml.

One of the real advantages offered by the use of the LAL test is the ability of the test to detect nonviable Gram-negative bacteria as well as viable organisms. Poor quality (high Gram-negative bacteria load) raw milk can be pasteurized, which destroys the bacteria but not the heat-stable enzymes.[12,13] This is often true in ultrahigh temperature (UHT) processed milk. Traditional cultural techniques are unsatisfactory in determining the keeping quality of these milks. Classical enzyme tests are cumbersome and typically not used in dairies. Since the LAL test detects both nonviable and viable Gram-negative bacteria, it provides a history of the bacterial milk quality from raw through pasteurized and may even be adapted to spray-dried material. Mikolajcik and Brucker[11] adapted the LAL test to predict keeping quality of milk by comparing LAL titers before and after incubation of the milk for 18 h at 13°C. Greater than a 2 $\log_{10}$ increase in endotoxin titer was observed to correlate with pasteurized milk that had poor keeping quality.

Various modification of the LAL test have been reported, including the use of a microdilution procedure.[14] The use of a chromogenic substrate which relies on the endotoxin-lysate enzyme complex releasing *p*-nitroanilide (*p*NA) from the synthetic peptide substrate Ac-Ile-Glu-Gly-Arg-*p*NA was reported by Iwanaga et al.,[15] and a modified microtiter test using one half of the LAL reagent was compared with the conventional method by May et al.[16] for the determination of Gram-negative bacteria in milk.

IV. CONCLUSION

As stated in the beginning of this chapter, there has been little research conducted on the uses of the LAL test in the veterinary sciences. The most notable findings have been the detection of mastitis in cattle and the evaluation of dairy products. However, many of the clinical uses of the LAL test pertaining to humans which are discussed in this book could be applied to similar veterinary situations. It is hoped that veterinary researchers will see new applications of the LAL test in veterinary medicine and begin evaluating this unique test where appropriate.

REFERENCES

1. **Hartman, I., Ziv, G., and Saran, A.,** Application of the *Limulus* amoebocyte lysate test to the detection of Gram-negative bacterial endotoxins in normal and mastitic milk, *Res. Vet. Sci.*, 20, 342, 1976.
2. **Schalm, O. W., Lasmanis, J., and Carroll, E. J.,** Pathogenesis of experimental coliform (*Aerobacter aerogenes*) mastitis in cattle, *Am. J. Vet. Res.*, 25, 75, 1964.
3. **Radostits, O. M.,** Coliform mastitis in cattle, *Can. Vet. J.*, 2, 401, 1961.
4. **Schalm, O. W., Carroll, E. J., and Jain, N. C.,** *Bovine Mastitis*, Lea & Febiger, Philadelphia, 1971.
5. **Levin, J., Tomasula, P. A., and Oser, R. S.,** Detection of endotoxin in human blood and demonstration of an inhibitor, *J. Lab. Clin. Med.*, 75, 903, 1970.
6. **Hakogi, E., Tamura, H., Tanaka, S., Kohata, A., Shimada, Y., and Tabuchi, K.,** Endotoxin levels in milk and plasma of mastitis-affected cows measured with a chromogenic Limulus test, *Vet. Microbiol.*, 20, 267, 1989.

7. **Thomas, B. S.,** Psychrotrophic micro-organisms in market cream. A review. I., *Dairy Ind.*, 35, 79, 1970.
8. **White, C. H., Gillis, W. T., Simmler, D. L., Galgl, M. K., Walsh, J. R., and Adams, J. T.,** Evaluation of raw milk quality tests, *J. Food Prot.*, 41, 356, 1978.
9. **Mikolajcik, E. M.,** Psychrotrophic bacteria and dairy product quality. I. Major organisms involved and defects produced, *Cult. Dairy Prod. J.*, 14, 6, 1979.
10. **Greer, G. G.,** Rapid detection of psychrotrophic bacteria in relation to retail beef quality, *J. Food Sci.*, 46, 1669, 1981.
11. **Mikolajcik, E. M. and Brucker, R. B.,** *Limulus* amoebocyte lysate assay. A rapid test for the assessment of raw and pasteurized milk quality, *Dairy Food Sanit.*, 3, 129, 1983.
12. **Law, B. A.,** Reviews of the progress of dairy science: enzymes of psychrotrophic bacteria and their effects on milk and milk products, *J. Dairy Res.*, 46, 573, 1979.
13. **Adams, D. M. and Brawley, T. G.,** Factors influencing the activity of heat-resistant lipase of *Pseudomonas*, *J. Food Sci.*, 46, 677, 1981.
14. **Prior, R. B. and Spagna, V. A.,** Adaptation of a microdilution procedure to the *Limulus* lysate assay for endotoxin, *J. Clin. Microbiol.*, 10, 394, 1979.
15. **Iwanaga, S., Morita, T., Harada, T., Nakamura, S., Niwa, M., Takada, K., Kimura, T., and Sakaibara, S.,** Chromogenic substrates for horeshoe crab clotting enzyme, *Haemostasis*, 7, 183, 1978.
16. **May, S. A., Mikolajcik, E. M., and Richter, E. R.,** Conventional tube and microplate *Limulus* amoebocyte lysate procedures for determination of Gram-negative bacteria in milk, *J. Dairy Sci.*, 72, 1137, 1989.

Chapter 15

EVALUATION OF STERILE PHARMACEUTICALS AND MEDICAL DEVICES — AN OVERVIEW

James F. Cooper

TABLE OF CONTENTS

I. Introduction 160
   A. Advent of Pharmaceutical Applications 160
   B. Sites of LAL Reagent Production 160
   C. Methods of LAL Reagent Production 160

II. Minimizing Variability in LAL Testing 160
   A. Reagent Related Sources of Variation 160
   B. Mechanisms of LAL Test Interferences 161
   C. Resolving LAL Test Interference Conditions 161
      1. Neutralization of pH 161
      2. Dispersion of Lipopolysaccharide Standards 161
      3. Selection of Standard Endotoxin Containers 162
      4. Maintenance of Critical Divalent Cation Concentrations 162
      5. LAL Reactive Material 163
      6. Serine Protease Materials 163
      7. Limitations of Quantitative Methods 163

III. Official Requirements 163
   A. Issues Addressed by Regulators 163
   B. Bacterial Endotoxins Test 163
   C. FDA Guideline for LAL Testing 164
      1. General Requirements 164
      2. Endotoxin Limits 164
      3. Testing for Product Interference 164
      4. Maximum Valid Dilution 165
      5. Test Methods for Human, Biological, and Veterinary Products 165
      6. Test Methods for Sterile Medical Devices 165

IV. Summary 165

References 166

I. INTRODUCTION

The most significant biomedical application of the *Limulus* amoebocyte lysate (LAL) test is evaluation of parenteral drugs and medical devices for endotoxin (lipopolysaccharide [LPS]) content. After 2 decades of use by the pharmaceutical industry, the LAL test is now the test of choice for certifying that a parenteral product is free of unsafe levels of bacterial endotoxin. In order that clinicians may benefit from the experience of industry, this discussion reviews differences in LAL production methods and reactivity, the causes of test variability and interference conditions, and key regulatory issues.

A. ADVENT OF PHARMACEUTICAL APPLICATIONS

The development of a viable alternative to the rabbit pyrogen test began with the innovative work of Johns Hopkins University investigators at the Woods Hole Oceanographic Institute. Levin and Bang showed that Gram-negative bacteria caused intravascular coagulation in the American horseshoe crab (*Limulus polyphemus*) and that the clotting phenomenon was due to an enzyme-mediated response to endotoxin by components in the intracellular fluid of the circulating cells, the amoebocytes.[1] The need for a suitable pyrogen test for radiopharmaceuticals led Cooper et al.[2,3] to extend this new approach to sterile pharmaceuticals. In the process, it was shown that LAL reactivity correlated well with pyrogenicity and that the new test was a more sensitive measure of endotoxin than the rabbit test by orders of magnitude.[2] Further, it was amenable to convenient, cost-effective testing for a variety of pharmaceutical products.

B. SITES OF LAL REAGENT PRODUCTION

Production sites for LAL are found at various locations along the eastern American coastline. (See Chapter 4 for a listing of licensed *Limulus* lysate suppliers.) Commercialization began in 1973 at two sites: Associates of Cape Cod, Inc., in Woods Hole, MA, under the direction of Watson and Sullivan, and Mallinckrodt, Inc., with its Chincoteague, VA facility under the direction of Mills and Cooper.

C. METHODS OF LAL REAGENT PRODUCTION

The desire to gain market advantage and governmental approval sparked a healthy competition in the LAL industry to improve the stability and sensitivity. Levin's[1] original LAL production method required collection of amoebocytes in a solution of 0.12% *N*-ethylmaleimide (NEM) and 3% NaCl (a solution designed to prevent degranulation of the cells) and subsequent washings to remove plasma components. Finally, the intracellular fluid was harvested by osmotic lysis, separated, and stored at 4°C. As Levin progressed in the development of a test for endotoxemia, the presence of inhibitors in human serum proved to be a perplexing problem. Suspecting the role of substances such as lipoproteins, he introduced a chloroform extraction of the LAL-serum sample to improve the sensitivity of the method.[4] Following this lead, Yin and Watson used chloroform extraction on crude lysate as a means of enhancing LAL reactivity by inhibitor removal.[5,6] The means used to achieve greater LAL sensitivity and integrity of gel formation vary markedly (Table 1), resulting in reagents that react differently under certain conditions.[6] All LAL suppliers use two or more of these modifications on their formulations to enhance sensitivity. Otherwise, the preparative method of Levin is still used today. Lyophilization is uniformly used to promote the stability and shelf life of lysate reagents (see also Chapter 3).

II. MINIMIZING VARIABILITY IN LAL TESTING

A. REAGENT RELATED SOURCES OF VARIATION

Clinicians must be aware that the heterogenous nature of LAL reagents leads to varied

TABLE 1
Methods Used to Enhance LAL Sensitivity

Aging of crude LAL
Adjustment of divalent cation concentration
Adjustment of coagulogen concentration
Addition of buffers and excipients
Addition of serum albumin and stabilizers
Chloroform extraction
Addition of biological detergents

responses in testing drug products, and that this variability surely extends to clinical and biological samples.[6] Further, LAL is optimized to detect Reference Standard Endotoxin (RSE), an *Escherichia coli*-derived lipopolysaccharide in water. Therefore, an alternate means of standardizing clinical material may be necessary in order for laboratories to compare their quantitative findings.

Reagent variability is quite apparent in testing drugs and devices.[6-8] A U.S. Food and Drug Administration (FDA) study[7] on drug compatibility with LAL revealed the extent of this source of test variation. Nine small-volume parenterals (SVPs) were tested with five licensed LAL reagents. Two lysates were found to have substantially more buffer capacity than others. It was significant that compatibility differed markedly (an eightfold dilution or greater) for seven of nine SVPs tested, even after pH adjustment. There is no clear advantage for any LAL supplier; rather, it is encouraging to know that if one reagent is not suitable in a given situation, another lysate may perform as expected. Test conditions that have unique interference problems may need to be matched with the most suitable reagent and test methodology.

B. MECHANISMS OF LAL TEST INTERFERENCES

Early acceptance of the LAL test was complicated because most products could not be tested directly. Interferences pose a challenge because the underlying mechanism is often difficult to identify. It is incorrect to assume that interference is always due to an effect of drug formulation or device extract on the LAL reagent. Rather, most interferences are due to an adverse effect of the test solution or container on the aggregation state of purified endotoxin controls. That is, inhibition is often a failure to recover inadequately dispersed LPS, a finding that may not represent the detection of naturally occurring endotoxin, such as that found in a clinical or biological sample.

Optimum LAL reactivity requires a neutral pH, availability of divalent cations and a dispersed endotoxin. Common interferences include extreme pH conditions, endotoxin modification, unbalanced cation levels, enzyme or protein modification, and nonspecific LAL activation. Table 2 summarizes interference mechanisms and their possible correction.

C. RESOLVING LAL TEST INTERFERENCE CONDITIONS

1. Neutralization of pH

The simplest solution to pH problems is dilution, taking advantage of the limited buffering capacity of LAL. Otherwise, one should initially add small, predetermined amounts of a dilute acid or base, and then use additional dilution if necessary.

2. Dispersion of Lipopolysaccharide Standards

It is recognized that vigorous mixing by vortex mixer is essential to disperse endotoxin standards. However, it is our experience that vortexing of biological samples containing natural endotoxin is unnecessary, causing unwanted foaming. Poor endotoxin recovery, observed as lost LAL activity, occurs because the lipophilic and hydrophilic regions of LPS permits interactions between other LPS molecules and certain test materials, resulting in micelle

TABLE 2
Corrective Steps for LAL Test Interferences

Problem	Solution
Suboptimal pH	1. Simple dilution 2. Addition of dilute acid or base
Endotoxin depletion	1. Simple dilution plus vortexing 2. Monitor/change containers 3. Use endotoxin dispersing agent
Calcium depletion	1. Dilute out chelation effect 2. Replace Ca and Mg deficits 3. Change LAL lot and/or supplier
Calcium potentiation	1. Change LAL lot and/or supplier
Container effects	1. Avoid polypropylene
Protein denaturation/ enzyme inactivation	1. Neutralize offending agent 2. Use ultrafiltration
Nonspecific activators	1. Neutralize offending agent 2. Change LAL lot and/or supplier

formation. The dispersibility of endotoxin is a function of the lipid A capability to form higher molecular weight aggregates, and of the electrostatic nature of cations associated with polysaccharide components of LPS. When aggregation increases, endotoxin dramatically decreases in both LAL reactivity and pyrogenicity. Generally, increased ionic strength produces endotoxin aggregation, particularly in concentrated monovalent cations and polyvalent anions. Endotoxin dispersing agents are often used to manage ionic strength-related inhibition. A 2% solution of a cationic dispersant (Pyrosperse, Whittaker Bioproducts) helps disperse endotoxin spikes in drugs containing oils, emulsions, and sodium salts of antibiotics.[7]

3. Selection of Standard Endotoxin Containers

The selection of container type and cleaning methods are keys to maintaining standard endotoxin controls. Endotoxin dilution tubes should be high-quality borosilicate glass to prevent adsorption of LPS to tube walls. Dilution tubes may be reused, but all traces of detergent must be removed to avoid micelle formation. The convenience of using plasticware is offset by unknown interference mechanisms that involve more than endotoxin adsorption. Tissue-culture grade polystyrene microplates and tubes may be used in contact with endotoxin solution, but complete loss of endotoxicity often occurs with storage in polypropylene tubes. Reproducible LAL testing necessitates vigilance to screen for and manage container-related problems.

4. Maintenance of Critical Divalent Cation Concentrations

Divalent cations are critical to LAL reactivity and dispersion of endotoxin. Reagents are optimized in the range of 0.01 to 0.05 *M* of divalent cations. LAL reactivity will be altered (suppressed or enhanced) if calcium and magnesium concentrations are not maintained. Organic chelators and anticoagulants (Na heparin, ethylenediaminetetraacetic acid [EDTA], and citrate) commonly found in clinical samples bind calcium by chelation and inhibit LAL reactivity. The best approach to correct cation depletion is to calculate the amount of calcium needed to neutralize the chelator, and replace the bound cations with calcium chloride. However, some LAL reagents are somewhat resistant to heparin inhibition.

5. LAL Reactive Material

Elutions or extracts from certain cellulose materials may contain traces of (1–3)-β-D-glucan, or LAL-reactive material (LAL-RM).[9,10] This entity apparently initiates the LAL gelation reaction by an alternate enzymatic pathway. The presence of LAL-RM has been reported in the blood of patients receiving dialysis by hollow-membrane fiber units or blood-product filtrates from cellulose acetate filters. This phenomena was associated with LAL processed by chloroform extraction.[9]

6. Serine Protease Materials

Chymopapain is a therapeutic protein, a serine protease, that can mimic endotoxicity. Trypsin is another protease that is LAL reactive under certain conditions. However, a heat treatment to denature the enzymes prior to LAL testing will readily destroy interferring proteases or protease inhibitors.

7. Limitations of Quantitative Methods

The chromogenic and kinetic-turbidimetric systems may offer the advantage of greater dilution to solve interference. However, gel-clot interference data rarely can be applied directly to these quantitative systems, necessitating new validation. For example, chromogenic assays introduce new problems as well as new opportunities, because the color procedure and reaction is quite different from the gel-clot reaction.[10] In kinetic-turbidimetric assays, the mechanisms of interference do not always affect the turbidity onset time as one might predict from gel endpoint data, so that comparison is not always meaningful. For example, a limitation of the LAL-4000 and LAL-5000 systems is that the LAL reagent is diluted by the specimen so that interference conditions are greatly exaggerated. Only the Toxinometer ET-201 permits simultaneous gel-clot and kinetic-turbidimetric data collection so that interference is identical.

III. OFFICIAL REQUIREMENTS

A. ISSUES ADDRESSED BY REGULATORS

Although the convenience and economy of the LAL test were recognized soon after its introduction, widespread use of this approach in the pharmaceutical/drug industry was delayed because the rabbit pyrogen test was an effective screening tool for endotoxin and was entrenched in drug approval systems and pharmacopeias. Before the *in vitro* approach could be substituted for the rabbit bioassay, government and standard setting organizations had to be convinced that the LAL test was sensitive, specific, reproducible, and amenable to standardization, and that endotoxin was the sole culprit in pyrogenic contamination. Years of experience with in-process testing of parenteral products by industry and surveillance testing by government agencies have shown that the LAL test is highly specific and that no significant nonendotoxin pyrogens occur in parenteral products that cannot be detected by other means. The LAL test is currently utilized throughout the pharmaceutical industry, with test applications ranging from monitoring pyroburden in raw materials to certifying the safety of a final product (Table 3).

B. BACTERIAL ENDOTOXINS TEST

Resolution of issues relating to sensitivity and standard methodology have proved to be time consuming. Standardization began with the appearance in 1980 of the Bacterial Endotoxins Test in the *U. S. Pharmacopeia XX (USP)*. The initial chapter focused attention upon a single methodology that is currently standardized only for gel-clot. Part of the standardization process required selection of a single endotoxin standard. Since the hydrophilic portion of the LPS molecule is highly variable among endotoxins, the proportion of lipid A differs widely, making comparison on a weight basis almost meaningless. The adoption of an *E. coli*-derived purified endotoxin characterized in endotoxin units (EU) provides a basis for comparing endotoxins that

TABLE 3
Pharmaceutical Applications of the LAL Test

Production water
In-process formulations
Biologics and blood products
Pharmaceutical products
Medical devices
Biotechnical products
Radioactive tracers
Tissue-culture media
Validation of dry-heat ovens

vary in specific biological activity. Since the RSE is expensive and exhaustible, the drug industry uses a Control Standard Endotoxin CSE, which is standardized in EU to the RSE for daily control studies. The CSE preparations are mostly *E. coli*-derived endotoxins. Clinical data should be reported in EUs for best comparison of data.

C. FDA GUIDELINE FOR LAL TESTING

1. General Requirements

In 1987, a guideline was released by the FDA to inform manufacturers of human drugs and biologicals, animal drugs, and medical devices of procedures the FDA considers necessary to validate the use of the LAL test as an end-product endotoxin test.[11] Those who adhere to the LAL guideline would be considered in compliance with relevant provisions of current Good Manufacturing Practices (cGMP) for drugs and devices and other applicable requirements.

2. Endotoxin Limits

The general endotoxin limit for parenteral drugs is 5 EU/kg dose, except for a 0.2 EU/kg limit for intrathecal drugs. Medical device eluates must not exceed 0.5 EU/ml; a 0.06 EU/ml limit applies to devices that contact cerebrospinal fluid.[3,11]

3. Testing for Product Interference

Before routine LAL testing is started, the potential for product inhibition must be determined and evaluated. Any pretreatment of a test product necessary to conveniently overcome interference needs to be validated using no less than three batches of each product. Inhibition exists if the endpoint of an assay of a twofold endotoxin dilution series made with the specimen (Positive Product Controls) differs more than one twofold dilution from the endpoint of a similar endotoxin series in water (Positive Water Control). For example, the following shows how inhibition might be recognized:

Labeled LAL Sensitivity (λ)	= 0.12 EU/ml
Endpoint Positive Water Controls	= 0.12 EU/ml
Endpoint Positive Product A Controls	= 0.20 EU/ml
Endpoint Positive Product B Controls	= 0.50 EU/ml

In the preceding example, Product A is considered within limits whereas Product B exhibits inhibition.

The easiest method to determine the noninhibitory product concentration is to prepare a series of increasing dilutions of the product containing a 2 λ endotoxin concentration.[6] Assay this series as well as a series of the product diluted with water. The following results are consistent with a product that is noninhibitory at a 1:20 dilution or greater, and is endotoxin-free.

Specimen Dilution	1:4	1:10	1:20	1:40
Product and 2 λ Endotoxin	– –	– +	+ +	+ +
Product and LAL Reagent Water	– –	– –	– –	– –

4. Maximum Valid Dilution

To avoid inhibitory test conditions and reduce the probability of detecting clinically insignificant endotoxin levels, the FDA allows specimen dilution based on the established endotoxin limits previously discussed. The USP has adopted specific limits for compendial items such as 175 EU per dose of radiopharmaceutical. These limits may be used to determine the extent of dilution that may be applied to overcome an interference problem without exceeding the limit endotoxin concentration. For drug products that have a published limit, the MVD may be calculated by the following formula:

$$\text{MVD} = \frac{\text{Endotoxin Limit} \times \text{Potency of Product}}{\text{Labeled Sensitivity}}$$

For example, the compendial limit for cyclophosphamide is 0.17 EU/mg. If an LAL reagent with $\lambda = 0.125$ is used to test this product where the potency is 20 mg/ml, the MVD equals 1:27 (the extent of permissible dilution to avoid inhibition).

$$\text{MVD} = \frac{0.17\ \text{EU/mg} \times 20\ \text{mg/ml}}{0.125\ \text{EU/ml}} = 27.2$$

5. Test Methods for Human, Biological, and Veterinary Products

The LAL test is validated for a pharmaceutical when three batches of a specific product are evaluated for compatibility. After confirmation of LAL reagent sensitivity, a positive product control series must have the same endpoint as the water controls. A new LAL reagent may be used after revalidating one lot of each preparation. Although most drugs are LAL tested, the industry is reluctant to apply LAL testing to biologics because LAL reactivity and pyrogenicity correlate poorly when measuring environmental endotoxin in biological samples.[12]

6. Test Methods for Sterile Medical Devices

Medical devices that are inserted into tissue, are part of a sterile surgical tray, or are labeled "pyrogen free" for any reason, must be evaluated by the LAL test. The limits for devices are concentration-based. Usually, 10 units are extracted or soaked in 400 ml of sterile water for injection for 1 h at ambient temperatures. This extraction may not contain more than 0.5 EU/ml.[11]

IV. SUMMARY

The LAL test has been accepted and standardized in the pharmaceutical and medical device industries because of its sensitivity, specificity, and reliability. In reviewing pharmaceutical applications, we find that LAL optimized for these products may not be ideal for clinical samples. A concern for clinicians is that LAL was standardized under somewhat artificial conditions, that is, the measurement of LPS in water. Rabbit pyrogenicity and LAL reactivity compare closely in this respect, but correlate poorly when measuring natural, environmental endotoxins in biological samples.[12] The latter finding should be a warning that current standardization of LAL may be inappropriate for comparing clinical data. A more suitable approach might be standardization of LAL using one or more clinically relevant, natural endotoxins suspended in human serum.

Current LAL reagents have been enhanced to the maximum with varied techniques and additives which produce unique reagent-related responses to certain test conditions. The pharmaceutical industry has developed clever ways to recognize and correct interference conditions in drugs and device extracts, but dilution is still the principal means to overcome inhibition. These corrective techniques are quite useful to clinicians, but loss of sensitivity by dilution may not be tolerable. It may be better for clinical samples to utilize less enhanced, less sensitive LAL that would be less responsive to agents such as divalent cation chelators (EDTA, heparin, etc.), endotoxin aggregation and LAL-reactive material.

The LAL test is the method of choice for monitoring endotoxin in all phases of pharmaceutical and device production, from raw material to finished product. A rigid set of regulations are in place to control LAL reagent preparation, standardization, and quality control of medical products. Although cumbersome, these regulations assure a continuous supply of sensitive, well-characterized and stable LAL reagents for use in industry and medicine.

REFERENCES

1. **Levin, J. and Bang, F. B.,** Clottable protein in Limulus: its localization and kinetics of its coagulation by endotoxin, *Thromb. Diath. Haemorrh.*, 19, 186, 1968.
2. **Cooper, J. F., Levin, J., and Wagner, H. N.,** Quantitative comparison of *in vitro* and *in vivo* methods for the detection of endotoxin, *J. Lab. Clin. Med.*, 78, 138, 1971.
3. **Cooper, J. F. and Harbert, J. C.,** Endotoxin as a cause of aseptic meningitis after radionuclide cisternography, *J. Nucl. Med.*, 16, 809, 1976.
4. **Levin, J., Tomasulo, P. A., and Oser, R. S.,** Detection of endotoxin in human blood and demonstration of an inhibitor, *J. Lab. Clin. Med.*, 75, 903, 1970.
5. **Novitsky, T.,** Discovery to commercialization: the blood of the horseshoe crab, *Oceanus*, 27(No. 1), 1984.
6. **Cooper, J. F.,** Ideal properties of a LAL reagent for pharmaceutical testing, in *Bacterial Endotoxins: Structure, Biomedical Significance, and Detection with the Limulus Amebocyte Lysate Test*, Ten Cate, J. W., Büller, H. R., Sturk, A., and Levin, J., Eds., Alan R. Liss, New York, 1985, 241.
7. **Twohy, C. W., Duran, A. P., and Munson, T. E.,** Endotoxin contamination of parenteral drugs and radiopharmaceuticals as determined by the LAL method, *J. Parenter. Sci. Technol.*, 30, 190, 1984.
8. **Steindler, K. A., Tsuji, K., and Enzinger, R. M.,** Potentiating effect of calcium gluconate on the Limulus Amebocyte Lysate (LAL) gelation-endpoint assay for endotoxin, *J. Parenter. Sci. Technol.*, 35, 242, 1981.
9. **Pearson, F. C., Bohon, J., Lee, L., Bruszer, G., Sagona, M., Morrison, D., and Dinarello, C.,** Comparison of chemical analyses of hollow-fiber dialyzer extracts, *Artif. Organs*, 8, 291, 1984.
10. **Ikemura, K., Ikegami, K., Shimazu, T., Yoshioka, T., and Sugimoto, T.,** False-positive result in Limulus test caused by *Limulus* amebocyte lysate-reactive material in immunoglobulin products, *J. Clin. Microbiol.*, 27, 1965, 1989.
11. Guideline on Validation of the LAL Test as an End-Product Endotoxin Test for Human and Animal Parenteral Drugs, Biological Products, & Medical Devices, U.S. Food and Drug Administration, 1987.
12. **Hochstein, H. D.,** The LAL test versus the rabbit pyrogen test for endotoxin detection: Update '87, *Pharm. Technol.*, 11, 124, 1987.

Chapter 16

PERSPECTIVES OF THE *LIMULUS* AMOEBOCYTE LYSATE TEST IN CLINICAL MEDICINE

Richard B. Prior

TABLE OF CONTENTS

I. Summary 168

II. Future Trends 170

References 172

I. SUMMARY

The *Limulus* amoebocyte lysate (LAL) test is the most sensitive method currently available to detect Gram-negative bacterial endotoxins. Endotoxins, complex lipopolysaccharides (LPS) found only in the cell wall of Gram-negative bacteria, are liberated into the surrounding environment upon bacterial lysis. They are widely distributed in nature, including the ground, air, and the water supplies,[1] and they are even found in mountain lakes and streams above 10,000 ft elevation in nonpopulated areas.[2] However, they are not found in significant levels, under conditions of good health, in the body fluids.[3] Endotoxins have both beneficial and detrimental effects upon humans, which were discussed in Chapter 1 by Miller and Hjelle. It is the detrimental effects and their unique association with Gram-negative bacteria that have prompted the development and use of LAL test in clinical investigations over past 25 years.

The LAL test was first described by Levin and Bang in 1964.[4] The key contributor to the LAL test is the horseshoe crab (*Limulus polyphemus*), a fantastic creature eloquently described by Shuster in Chapter 2. The basis for the LAL test is an enzymatic reaction which occurs between endotoxin and lysate made from the amoebocytes of the horseshoe crab and leads to *in vitro* gelation (Chapter 3). Many modifications have since been made to improve the sensitivity of the LAL test, including the addition of a chromogenic substrate. Picrogram quantities (10^{-12} g) of endotoxin can now be detected quickly and accurately with a minimum of time (1 h, gel; 10 min, chromogenic substrate), laboratory equipment, and technical expertise. The quality of lysate and its approved uses in clinical medicine are regulated by the U.S. Food and Drug Administration (FDA) (Chapter 4). Lysate can be obtained commercially from several different manufacturers in bulk form (multitest vials) or in single-test vials. A chromogenic substrate procedure and two automated turbidimetric methods are also licensed and approved (Chapter 4).

While the LAL test is simple to perform, certain precautions must be taken to ensure accurate test results. For example, since endotoxins are widely distributed in nature, great care must be exercised when collecting, handling, and processing samples for LAL testing. A test result is only as good as the specimen submitted for testing. The presence of interfering substances in the test sample must also be determined and appropriate controls included in the test procedure (Chapters 3, 15).

The LAL test has been used most successfully in the pharmaceutical industry, where rapid and economical testing of parenteral fluids, biologicals, and invasive medical devices for contaminating "pyrogens" is essential (Chapter 15). However, the clinical utility of the LAL test appears to depend upon the specific clinical situation being evaluated. The various chapters in this book have detailed the uses of the LAL test in specific clinical settings.

In Chapter 5, Pearson discussed the detection of endotoxemia with the LAL test. This was one of the earliest clinical situations evaluated with the LAL test and one of the most controversial. Some investigators demonstrated clinical utility while others reported the lack of clinical utility. Detecting endotoxins in blood can be very difficult and interpreting the results can be equally difficult. However, if the blood is prepared properly to remove the nonspecific activators and inhibitors, accurate and meaningful results can be obtained. In appears that collecting whole blood in 15 to 30 U final concentration of nonpyrogenic heparin, removing and diluting the platelet-rich plasma 1:10 in pyrogen-free water, and subsequently heating the diluted plasma at 70°C for 10 min or 75°C for 15 min will produce the most accurate and reproducible results. The addition of a chromogenic substrate has also greatly improved the LAL test in this clinical setting. It now appears that some clear demarcation of endotoxin concentration in blood can be made between healthy humans and patients with Gram-negative sepsis. Some form of standardization is now needed for the detection of endotoxins in blood and clinical trials must be run to determine unequivocally the clinical utility of the LAL test.

Two of the most successful indications of the LAL test in clinical medicine have been its use in evaluating meningitis and urinary tract infections (UTI). Nachum discussed and summarized the numerous articles published on the uses of the LAL test in these two clinical settings (Chapters 6 and 7). It is clear that the LAL test does have clinical utility in managing cases of meningitis. The test is very sensitive and specific in quickly detecting Gram-negative bacterial involvement. Additionally, the inhibitors found in blood are not present in nonsanguineous cerebrospinal fluid (CSF), and pretreatment of the clinical sample is unnecessary. CSF can be tested at the time of the spinal tap at the patient's beside, and involvement of Gram-negative bacteria can be determined in 10 min or less if a chromogenic substrate is used. The LAL test is superior to the Gram stain in detecting Gram-negative bacteria and can provide useful information for patients who received antibiotic therapy before testing. In these cases, Gram-stained smears and cultures are usually negative but the LAL tests are positive since the test detects endotoxin and not viability of the organism. Thus, it is important to determine whether antibiotics have been previously administered when interpreting positive LAL test results in patients with negative microbiological findings for Gram-negative bacteria.[3]

The use of the LAL test in the detection of significant ($\geq 10^5$) Gram-negative bacteriuria appears promising and test kits could be developed which would be amenable for use in a private practice setting. The test could also be automated and used for large-scale urine screening in hospitals. However, a negative LAL test will not rule out the possibility of less than 10^5 Gram-negative bacteria per milliliter of midstream urine or other infectious agents lacking endotoxins.

The utility of the LAL test in the evaluation of exudative urethritis in males has been well documented in numerous studies summarized in Chapter 7. The test was shown to be fast, simple to perform, accurate, and clinically useful in this setting. The FDA did approve the test for the rapid presumptive diagnosis of gonococcal urethritis in males and a test kit was patented in the U.S. and in other countries. However, the test kit (Gonoscreen) is, at present, not available pending its reintroduction to the market at a later date. The test was subsequently improved to include urogenital swabs for sample collection and a chromogenic substrate for visible color endpoint determinations. Use of the swabs increased by approximately 30% the number of evaluable patients and the use of a chromogenic substrate reduced the testing time from 30 to 10 min. Thus, use of the LAL test in a private practice setting where Gram staining facilities may not be generally available would provide test results equal to Gram-stain smears read by experienced microscopists and to cultures processed in sexually transmitted disease (STD) clinics. Additionally, test results would be available at the time of the patient's initial evaluation. Appropriate therapy could then be initiated immediately and a sound, informed disposition of sexual contacts could be contemplated.

The usefulness of the LAL test in evaluating females for gonococcal cervicitis lies in its ability to exclude this disease because of its high negative predictive value (100% regardless of the prevalence rate of cervical gonorrhea). Exclusion of STDs is just as important as the diagnosis of such diseases. Therefore, the LAL test does have potential clinical utility as a screening test to exclude gonococcal cervicitis and also to determine whether sexual contacts to patients with gonorrhea should be treated prophylactically with antibiotics (epidemiological therapy). However, additional studies are needed to substantiate these preliminary findings in females.

In Chapter 9, Elin discussed the role of the LAL test in evaluating pyogenic arthritis. Only four studies have been published assessing the utility of the LAL test in this clinical setting. In one study, the test seemed to lack specificity for a septic process, but a negative LAL test result indicated nonseptic joint disease. In another study, the LAL test appeared to have specificity but lacked sensitivity. For diagnosing septic arthritis due to Gram-negative bacteria, the sensitivity and specificity of the LAL test was reported to be 82 and 80%, respectively, using current methodologies and culturing techniques. Elin suggests that studies which optimize specimen collection and treatment of serum prior to testing with state-of-the-art LAL technology are needed to define the clinical value of the test in evaluating pyogenic arthritis.

The use of the LAL test in evaluating otitis media with effusion (OME) (Chapter 10) appears to have potential clinical utility and may also aid in the understanding of the inflammatory responses seen in middle ear infections. Additionally, the LAL test may prove helpful in the assessment of the ability of endotoxin to pass from the middle to the inner ear, presumably via the round window membrane. Obviously, more clinical research is needed to better assess the role of the LAL test in studying and evaluating ear infections.

Detection of ocular infections with the LAL test looks very promising and Alfonso and Miller discussed the results they have obtained at the Bascom Palmer Eye Institute (BPEI) over the past 10 years (Chapter 11). The LAL test correlated very well with conventional microbiological methods used to evaluate keratitis and endophthalmitis, and its use could help guide the physician in the selection and initiation of an early, specific, and appropriate therapy. In addition, these investigators have been using the test to detect contaminating endotoxins in ophthalmic solutions and topical ocular medications. The LAL test may offer an early, rapid, and reliable warning system for reducing and avoiding the ulcerative keratitis associated with contact lens wear. Other ocular infections (keratitis, endophthalmitis, sclerokeratitis, and conjunctivitis) associated with patient-contaminated ocular medications may also be reduced or prevented.

The use of the LAL test in monitoring patients on continuous ambulatory peritoneal dialysis (CAPD) was discussed in Chapter 12 by Smalley. Sensitivity to detect peritonitis caused by Gram-negative bacteria was reported to be 100% for CAPD patients. The rapid detection of Gram-negative associated endotoxin by the LAL test may aid in early initiation of appropriate antibiotic therapy in these patients. Additionally, CAPD patients may be able to monitor their dialysate at home using a simplified chromogenic substrate LAL test kit. The LAL test also shows potential in detecting gastrointestinal (GI) perforation following blunt abdominal trauma (Chapter 12). Additionally, the presence of circulating endotoxins may reveal perforation following certain diseases or infections of the GI tract. Only a few studies have been done to evaluate the LAL test in these clinical settings. While the preliminary results look very promising, more studies are needed to confirm the clinical utility of the test.

The allied medical uses of the LAL test have been numerous and varied. Pearson discussed the more important and relevant ones in Chapter 13. Monitoring hemodialyzers and the water used to prepare the disinfectant for reuse with the LAL test is now standard in most places. In fact, the Association for the Advancement of Medical Instrumentation (AAMI) guidelines for reuse recommend that water be tested at least monthly by the LAL test and should contain less than 1 ng of endotoxin per milliliter. Surveillance of respiratory equipment for contamination with Gram-negative bacteria by the LAL test has also been reported as showing good correlation with culture results. In addition, the detection of contaminated foods and water has been evaluated and results were very favorable. What is now needed, especially in these situations where there is a wide variation between types of testing material, is some form of standardization.

There has been little research conducted on the uses of the LAL test in the veterinary sciences. The most notable have been the detection of mastitis in cattle and the evaluation of dairy products which Richter and Mikolajcik summarized in Chapter 14. However, many of the clinical investigations of the LAL test pertaining to humans which are discussed in this book could be applied to similar veterinary situations. Perhaps veterinary researchers will see new applications of the LAL test in veterinary medicine and begin evaluating the test where applicable.

II. FUTURE TRENDS

Where will the LAL test be 25 years from now? Based on the indications of the past 25 years, the test will still be a valuable tool for research and be used in clinical cases where Gram-negative infection or endotoxin contamination is a problem. The LAL test is an antigen detection test similar to the Gram stain in that the organism is detected. Nothing can be said about the organism, apart from morphology, other than it is Gram negative — genus and species cannot be determined. However, unlike the Gram stain, it also detects fragments of the organism (free endotoxin), and it is more sensitive in most cases. Moreover, the LAL test is much easier to perform and interpret. During this author's experience with the LAL test, there has been some criticism since the test does not involve an "antigen-antibody" reaction in the classical sense. In a way, this is an advantage for the test rather than a detriment. The detection of a homologous core (lipid A) which is common among the Gram-negative bacteria rather than the detection of a species specific antigen can be desirable, especially when the therapeutic choices are based on the presence or absence of a class of microorganism, i.e., Gram-negative bacteria.

Efforts have been directed toward improving the sensitivity of the LAL test. This has been accomplished primarily by using a chromogenic substrate. The newer amplification method reported by Roth et al.[5] looks very interesting and shows that newer ways and approaches to detect endotoxin-initiated chemical products are areas for further research. The novel modification of the LAL test by Zhang et al.[6] looks very promising and also points the way to further areas of investigation. These researchers found that coagulogen losses antigenicity during the LAL reaction and that this can be detected easily by using an enzyme-linked immunosorbent assay (ELISA) in conjunction with the LAL test. They reported a sensitivity down to 0.1 pg/ml of test sample. It appears that the level of sensitivity is low enough now to be clinical useful in cases where extreme sensitivity is required, i.e., endotoxemia. Certain clinical applications, however, require less sensitive lysate. For example, in the evaluation of exudative urethritis in males, a lysate with a sensitivity of only 0.625 to 2.5 endotoxin units (EU)/ml is required for optimal performance of the test since the gonococci produce large quantities of endotoxins.

Efforts should also continue to improve the specificity of the LAL test in certain settings. The alternate (1–3)-β-D-glucan pathway discussed in Chapter 3 can cause nonendotoxin "false" positive LAL test results. A new chromogenic LAL test (Endospecy) has been modified so that (1–3)-β-D-glucan cannot activate the *Limulus* lysate cascade and the result is a chromogenic endotoxin-specific LAL test.[7] Perhaps the amoebocytes could be cultured *in vitro* or the gene that codes for the proclotting enzyme could someday be identified and transferred into a bacterium. Then large quantities of endotoxin-reactive material could be made and the need to harvest and bleed the horseshoe crabs would be eliminated. Chemical modifications of the pure enzyme may then lead to the development of "lysate" which is consistent, sensitive, and specific. This type of effort will no doubt be very expensive, and it remains to be seen whether the clinical needs will justify such an effort. Therefore, at the present time, clinical research involving the LAL test will be conducted using lysate preparations which are commercially available and used primarily in the pharmaceutical industry.

Improvements in further reducing incubation time would be clinically advantageous in some cases, and elimination of the need for an incubator would also be useful. Developing LAL tests which incorporate the reagents in paper strips or absorbant material much like the "dip stick" or ELISA-filtration techniques would further simplify the test. Improvements in the imperfect cultural standards to which the LAL test results are compared should also be made. With the advent of the noncultural antigen detection methods used to identify specific infecting microorganisms (DNA probes, ELISA, etc.), research protocols using the LAL test should also include these tests in addition to conventional cultures for more meaningful data analysis.

A fascinating new area in clinical medicine involves the use of monoclonal antibodies directed toward endotoxins to protect against Gram-negative shock. Large clinical trials are currently underway to examine the clinical utility of these antibodies in preventing septic shock. A preliminary report suggested the failure of the LAL test to predict the *in vivo* efficacy of these anti-endotoxin monoclonal antibodies.[8] Since the dynamics of endotoxins in and their clearance from the circulatory system are so complex, additional studies are needed, perhaps using controlled animal models, to confirm these findings. A good animal model for the study of endotoxins in blood would also be useful for studying the clinical efficacy of the LAL test in evaluating endotoxemia under controlled conditions.

Another exciting new area where the LAL test potentially may have clinical utility is in organ transplantations. Yokoyama et al.[9] have recently found a striking association between perioperative blood endotoxin levels, the difficulty of convalescence, and the ultimate outcome. In the study of 90 liver transplantations in adult humans, these investigators found that the presence of high endotoxin levels preoperatively and at the end of the anhepatic period was associated with graft failure and a high mortality. Patients with primary nonfunction of their transplants typically had severe endotoxemia. They concluded that endotoxemia could be a cause rather than an effect of perioperative complications and graft loss.[9] Perhaps studies using the anti-endotoxin monoclonal antibodies in transplant patients with elevated perioperative blood endotoxin levels could be undertaken to determine whether such therapy would improve convalescence and the ultimate outcome.

It appears that endotoxin in clinical samples reacts with *Limulus* lysate differently than purified control standards. Cooper discussed the differences between clinical (native) and laboratory endotoxins and suggested that clinicians be aware of these when conducting clinical research with this test (Chapter 15). He also suggested that the current methods of standardizing the LAL test in water may not be applicable to clinical material and perhaps a more suitable approach might be standardization of the LAL test using one or more clinically relevant, natural endotoxins suspended in human serum.

Lysate producers should work closely with the clinical investigators and modify lysate formulations where appropriate. They could also be instrumental in helping standardize the test to perform better in certain clinical situations. Collaboration between producers, users, and researchers is the key. Perhaps more relevant research studies covering all aspects of the clinical uses of the LAL test could be conducted in a forum of mutual cooperation and sharing of scientific ideas. One such forum could be the International Endotoxin Society and its annual Congress devoted to endotoxin research. Periodic symposia on the clinical uses of the LAL test would also be beneficial.

Finally, none of the clinical applications described in this book has replaced and probably never will replace the conventional methods used in these settings. However, with a thorough understanding of both the advantages and the limitations of the LAL test, it can serve as an extremely valuable adjunct to other diagnostic modalities in the evaluation of certain Gram-negative infections.

REFERENCES

1. **DiLuzio, N. R. and Friedmann, T. J.,** Bacterial endotoxins in the environment, *Nature*, 244, 49, 1973.
2. **Prior, R. B.,** unpublished data, 1983.
3. **Prior, R. B.,** Bacterial endotoxin detection by the *Limulus* amoebocyte lysate test, in *Antigen Detection to Diagnose Bacterial Infections*, Vol. 2, Kohler, R. B., Ed., CRC Press, Boca Raton, FL, 1987, chap. 30.
4. **Levin, J. and Bang, F. B.,** The role of endotoxin in the extracellular coagulation of *Limulus* blood, *Bull. Johns Hopkins Hosp.*, 115, 265, 1964.
5. **Roth, R. I., Levin, J., and Behr, S.,** A modified *Limulus* amebocyte lysate test with increased sensitivity for detection of bacterial endotoxin, *J. Lab. Clin. Med.*, 114, 306, 1989.
6. **Zhang, G.-H., Baek, L., and Koch, C.,** New microassay for quantitation of endotoxin using *Limulus* amebocyte lysate combined with enzyme-linked immunosorbent assay, *J. Clin. Microbiol.*, 26, 1464, 1988.
7. **Obayashi, T., Tamura, H., Tanaka, S., Ohki, M., Takahashi, S., Arai, M., Masuda, M., and Kawai, T.,** A new chromogenic endotoxin-specific assay using recombined *Limulus* coagulation enzymes and its clinical applications, *Clin. Chim. Acta*, 149, 55, 1985.
8. **McConnell, J. S., Appelmelk, B. J., and Cohen, J.,** Failure of the LAL assay to predict *in vivo* efficacy of anti-LPS monoclonal antibodies (Mabs), abstract #1039 presented at the 28th Interscience Conference on Antimicrobial Agents and Chemotherapy, October 23 to 26, 1988, Los Angeles.
9. **Yokoyama, I., Todo, S., Miyata, T., Selby, R., Tzakis, A. G., and Starzl, T. E.,** Endotoxemia and human liver transplatation, *Transplant. Proc.*, 21, 3833, 1989.

INDEX

A

Abdominal trauma, blunt, 137—138
Accuracy of test, see also Sensitivity of assays; Specificity of assays; specific applications
  in endotoxemia, 56—57
  in eye infection, 130
Achromobacter, 156
Acinetobacter lwoffi, 76
Adjuvancy, 8
Adult respiratory distress syndrome, 55
Aeromonas, 156
Aggregation state, and pyrogenicity, 141—142
Airborne endotoxin, 148
Alcalgenes, 156
Ammonium sulfate precipitation, 58
Amoebocytes, 19—21, 28
Analogs of lipid A, 4
Antibodies, passive immunization with, 53
Anticoagulant property, 9
Antigen detection methods, 53
  chemical structure of lipopolysaccharides, 4—7
  in meningitis, 70, 77
Antisera, 53
Antithrombin III, 56, 59
Antitrypsin, 56
Arthritis, pyogenic, 112—114
Aspergillus species, 129
Automated systems
  FDA regulation, 46—47
  methods, 32—33

B

Bacillus, eye infection, 128
Bacillus cereus, eye infection, 126
Bacteriuria, 82—93, see also Urine assays
Bacteroides fragilis, peritonitis, 137
Biological activity of endotoxin, 3
Biology of *Limulus polyphemus*, 16—24
  body sections and features, 16—18
  environmental and biological interfaces, 24
  hemolymph, 19—21
  life cycle, 21—22
  major organ systems, 18—19
Blepharitis, 122
Blunt abdominal trauma, 137—138
Bradykinin, 52
Brucella, 8

C

C3H/HeJ model, 6, 8
Cachectin, 9
Canaliculitis, 122
Candida, eye infections, 125, 128, 129
Capillary-leak syndrome, 53
Carbohydrate metabolism, 9
Carrageenan, 57
Cellulitis, 122
Cerebrospinal fluid (CSF)
  C-reactive protein in, 55
  meningitis, 68—71
Cervical exudates, 55
Cervicitis, gonococcal, 105—106
Chemical structure of lipopolysaccharides, 4—7
Chemical synthesis, 4
Chlamydia trachomatis, 98, 100
Chloroform extraction, 59
Chromogenic substrate method
  FDA regulation, 46
  methods, 31—32
Citrobacter diversus, eye infection, 129
Citrobacter freundii, peritonitis, 137
Coagglutination, in meningitis, 77
Coagulation mechanisms, 52
Coagulation pathways, 10, 52
Colony stimulating factors (CSF), 7, 9, 55
Complement, 52, 53
Conjunctivitis, 122, 123
Contact lens solutions, 130—131
Continuous ambulatory peritoneal dialysis, 136—137
Core polysaccharide
  chemical structure, 5—6
  monoclonal antibody, 53
Cornea, keratitis, 122—126
Corticosteroids, 10
Corynetacterium, eye infections, 125, 129
Countercurrent immunoelectrophoresis, 70, 77
C-reactive protein, 55
CSF, see Colony stimulating factors and Cerebrospinal fluid
Curvularia sp., eye infection, 125, 129
Cyclooxygenase, 9
Cytokines, 7, 8, 10

D

Dacryocystitis, 122
Delipidation of serum, 56
Demography, gonorrhea, 99
Dextrans, 57
Diagnosis, see Endotoxemia; specific disorders
Dialysis
  continuous ambulatory peritoneal (CAPD), 140—141
  hemodialysis, 140—147
DIC, see Disseminated intravascular coagulation
Dilution, 57—58
Diphtheria antitoxins, 8
Disseminated gonorrhea, 106—107
Disseminated intravascular coagulation (DIC), 8, 52—55
Dohle bodies, 55
Dry up method, 30
Dyes, 30

E

Ear infection, 116—119
Edema, 9
Efficacy, in meningitis, 74—77
Elastase-α[1]-proteinase, 69—70
ELISAs, see Enzyme-linked immunosorbent assays
Endophthalmitis, 122, 123, 126—128
Endorphins, 9
Endospecy, 32, 73—74
Endothelial cells, 9
Endotoxemia, 10
  clinical diagnosis, 55
  clinical syndrome, 53—54
  laboratory diagnosis, 55—56
  sample preparation, 56—60
    ammonium sulfate precipitation, 58
    chloroform extraction, 59
    ether method, 58
    false negatives, 56—57
    false positives, 57
    heat and dilution, 57—58
    perchloric acid, 59
    pH shift method, 59—60
    platelet extraction, 58
    resin extraction, 59
Endotoxins, 2—11
  aggregation state, and pyrogenicity, 141—142
  beneficial effects, 7—9
  chemical structure of lipopolysaccharides, 4—7
    core polysaccharide, 5—6
    lipid A, 6—7
    O-antigen, 4—5
  detrimental effects, 9—10
  FDA regulation, 40—43
  LAL test and, 10—11
  medical equipment surveillance
    hemodialysis, 140—147
    respiratory, 148—149
Enterobacter, 76
Enterobacter cloacae peritonitis, 137
Enterobacteriaceae, 76
Enzyme-linked immunosorbent assays (ELISAs), 3, 31, 61
  in gonorrhea, 100
  in meningitis, 70
Enzyme immunoassays, 70, see also ELISAs
Epidemiology, gonorrhea, 98—99
Escherichia coli
  hemodialysis, 141
  meningitis, 68, 72, 76
  milk, 156
  septic shock, 52
Ether method, 58
Eye infections, see Ocular infections

F

Factor C, 32
Factor G, 32, 73
Factor X, 59
Factor XII, 56
False-negative Gram stains, 68—69
False-negative *Limulus* tests, 30, see also Sensitivity of assays
  in cervicitis, 105—106
  in endotoxemia, 56—57
  in meningitis, 75—76
  in urethritis, 102—105
  urine screening, 88—89
False-positive *Limulus* tests, see also Specificity of assays
  in cervicitis, 105—106
  in endotoxemia, 57
  in meningitis, 74—75
  in urethritis, 102—105
  urine screening, 89—90
FDA, see Food and Drug Administration
FDA guidelines, 43—45, 164—165, see also Regulation, federal
Federal register notices, 43—45
Federal regulation, see Regulation, federal
Fever production, 8
Fibrin degradation products, 55
Fibrinogen levels, 55
Flavobacterium
  contamination of humidifier water, 148
  milk, 156
Flocculation, 74
Food and Drug Administration (FDA), see Regulation, federal
Food contamination, 149—150
Freeze-dried lysate, 28
Fungi, eye infections, 123, 125, 129
Fusarium, 123, 125, 129

G

Gel end point method
  FDA regulation, 46
  methods, 29—31
Glassware, 30
Gonorrhea, 98—107
  cervicitis, 105—106
  evaluation of sexual contacts and disseminated gonorrhea, 106—107
  urethritis, 98—100
    demography and clinical characteristics of urethritis, 99
    diagnosis in males, 100—105
    epidemiology of urethritis, 98—99
    Gram stain, 99—100
    noncultural methods to detect gonorrhea, 100
    test kit for, 102—104
Government regulation, see Regulation, federal
Gram-negative bacteria, 76, see also specific pathogens
  anti-endotoxins, 7—8
  septicemia/shock, 7, 9, 10, see also Endotoxemia
Gram stain
  gonorrhea, 99—100
  meningitis, 68, 71, 76—77

Group B streptococci
  eye infections, 123, 125, 128, 129
  meningitis, 68, 70
Guidelines, see Regulation, federal

H

Haemophilus influenzae, 122
  eye infections, 122, 123, 125, 128, 129
  meningitis, 68, 70, 72, 76, 77
  otitis media, 116
Head injuries, 69
Heat, 57—58
Hemodialyzers, 140—147
Hemodynamic alterations, 55
Hemolymph, 19—21
High density lipoproteins, 56
Historical background, 28
Horseshoe crab, see *Limulus polyphemus*
Humidifier fever, 148
Humoral immune responses, 7
Hyperimmune serum, 7
Hypoglycemia, 9
Hypotension, 8, 9
Hypovolemia, 9

I

Immunomodulatory effects, endotoxin, 8
Interference, pharmaceuticals, 161—163
γ-Interferon, 7, 9, 54
Interleukin-1, 10, 52, 142—144
Interleukin-2, 9
Interleukin-6, 9
Intestinal endotoxin, 52

K

Keratitis, 122—126, 130
Kinetic turbimetric assay, 46—47
Kinin, 52
Klebsiella, 76
Klebsiella pneumoniae
  meningitis, 76
  peritonitis, 137
  septic shock, 52
Kupffer cells, 8, 52

L

Labeling, FDA regulation, 47—48
Laboratory diagnosis, see also specific clinical applications
Lacrimal sac, 122
LAL, see *Limulus* amoebocyte lysate
LAL-5000 assay, 32, 46—47
LAP, see Lipid A-associated protein
Latex particle agglutination, meningitis, 77
Licensing, FDA
  manufacturers, 45
  test, 39—40
Limulus amoebocyte lysate test (LAL), 28—34
  historical background, 28
  lysate preparation, 28—29
  mechanism, 29
  methods, 29—33
    automated systems, 32—33
    chromogenic substrate, 31—32
    gel end point, 29—31
    other, 33
    role of FDA in regulating test, see Regulation, federal
    specificity of, 29—32
      β-glucan activation, 32—33, 147
    turbidimetric, 31
Limulus polyphemus, biology of
  body sections and features, 16—18
  environmental and biological interfaces, 24
  hemolymph, 19—21
  life cycle, 21—22
  major organ systems, 18—19
Lipid A, 4, 10
  chemical structure, 6—7
  interferon induction, 9
Lipid A-associated protein (LAP), 6
Lipopolysaccharides (LPS), chemical structure, 4—7
Lipoteichoic acid, 7
Listeria monocytogenes
  eye infection, 122
  meningitis, 68, 76
Liver, 8
Liver dysfunction, 57
LPS, see Lipopolysaccharides
Lyophilized lysate, 28
Lysate preparation, 28—29

M

Male *Limulus polyphemus,* 23
Mannans, 57
Mastitis, 156
Medical applications, nonclinical, 140—150, see also Pharmaceuticals
  contaminated foods and water, 149—150
  hemodialysis
    dialysate, 145—146
    endotoxin evaluation of hemodialyzer, 146—147
    evaluation of hemodialyzers and reuse water, 140—147
    generation of interleukin-1 by endotoxin, 142
    interleukin-1, 142—144
    membrane permeation of endotoxin, 141—142
    pyrogen evaluation of water in hemodialysis, 144
    pyrogenic reactions in hemodialysis, 141
    treatment, 140—141
    water quality standards for hemodialysis reuse, 144—145
  respiratory equipment surveillance, 148—149
Membranes, hemodialyzer, 141—142, 146—147

Meningitis, 68—78
  efficacy of screening, 74—77
  laboratory diagnosis, 68—70
  literature review, 70—77
Methods, 29—33, see also specific applications
  FDA approved, 46
  sample preparation in endotoxemia, 56—60
Microdilution plates, 30
Microslides, 30
Middle ear infections, 116—119
Milk quality, 156—157
Mitogenic factor in LPS, 8
Mitogenicity, endotoxin, 8
Monoclonal antibody, 53
Moraxella (Branhamella) catarrhalis, 116, 125, 129
Morganella morganii, 125, 129
Mortality rates, sepsis, 52
Multiple organ failure, 10
Multi-test vials, 28
Muramidase, 69
Mycobacterium
  eye infection, 126
  meningitis, 76

N

Neisseria gonorrhoeae, see also Gonorrhea
  cervicitis, 105—106
  eye infection, 122, 123, 125, 129
  urethritis, 98—100
Neisseria meningitidis
  eye infection, 122
  meningitis, 68, 70, 72, 76, 77
Neonatal meningitis, 68, 71—72
Neoplasms, 69
Neutrophilic changes, 55
Neutrophils, 9
p-Nitroanilide (PNA) group, 31

O

O-antigen, chemical structure, 4—5
Ocular infections, 122—131
  endophthalmitis, 126—128
  keratitis, 122—125
  medications, 131
  ophthalmic solutions, 130—131
  ophthalmology, 128—130
Opioids, 9
Orbital cellulitis, 122
Organ transplants, 173—174
Otitis media, 116—119
Oxygen intermediates, 9

P

Passive immunization, 53
Pathophysiology, endotoxin, 3
  beneficial effects, 9—10
  detrimental effects, 7—9
Peptidoglycan, 57
Perchloric acid, 59
Peritonitis, 136—138
  peritoneal dialysis and, 136—137
  trauma and, 137—138
Pharmaceuticals, 160—166
  minimizing variability in testing, 160—163
  official requirements, 163—165
  ophthmalic solutions and medications, 130—131
  production, 160
Phosphoenolpyruvate carboxykinase, 9
pH shift method, 59—60
Plasma proteins, 58
Platelet-activating factor, 10
Platelet extraction, 58
Platelets, 56
Polycystic kidney disease (PKD), 10, 84—85
Polymorphonuclear leukocytes, 55
Predictive value of LAL test
  in gonococcal cervicitis, 106
  in gonococcal urethritis, 105
  in urinary tract infection, 86—88
Procoagulant activity, 10
Propionibacterium acnes, 126
Protein assays, 30
Proteus, 52
Proteus mirabilis, eye infections, 123, 125, 129
Prothrombin times, 55
Pseudomonas, 6
  meningitis, 76
  milk, 156
Pseudomonas aeruginosa
  dialysate, 146
  eye infections, 122, 123, 125, 126, 129
    eye medication, 131
    ophthalmic solutions, 130—131
Pseudomonas alcaligenes, in peritonitis, 137
Pseudomonas diminuta, 6
Pseudomonas vesicularis, 6
Purulent meningitis, 68
Pyogenic arthritis, 112—114
Pyrogenicity, 8
  aggregation state and, 141—142
  hemodialysis, 141, 144

Q

Quantitative test, 31
Quellung test, meningitis, 70

R

Radiation detoxified LPS, 8
Radioisotopes, 30
Regulation, federal, 38—49
  automated systems approved by FDA, 46—47
    LAL-5000 and kinetic turbidimetric assay, 46—47
    toxinometer ET-201, 47
  background of testing in FDA, 38—39
  FDA's experience with endotoxin, 40—43
  Federal Register notices, 43—45
  first uses of LAL test, 39
  labeling, 47—48

legal effect of guideline, 45
licensed manufacturers, 45
licensing of test, 39—40
methods, FDA-approved, 46
chromogenic substrate method, 46
gel end point method, 46
turbidimetric method, 46
pharmaceutical application, 163—165
testing U.S. standard endotoxin in rabbits and humans, 48
Renal failure, see Dialysis
Resin extraction, 59
Respiratory equipment surveillance, 148—149
Reticuloendothelial system, 7

S

Salmonella minnesota, 5
Sample preparation, 56—60
Screening, meningitis, 74—77
Sensitivity of assays, 29, 30, 61
in endotoxemia, 56—57
in eye infection, 130
in meningitis, 74, 75—76
in urethritis, 98—100
urine screening, 88—89
Sepsis, 10, see also Endotoxemia
Serratia, 76
Serratia marcescens
eye infections, 122, 123, 125, 129
eye medication, 131
ophthalmic solutions, 130, 131
meningitis, 72, 76
Serum-associated LPS, 6
Sexually transmitted diseases, see Gonorrhea
Shock, 9, see also Endotoxemia
Specificity of assays
in endotoxemia, 57
in eye infection, 130
in meningitis, 74—75
in urethritis, 98—100
urine screening, 89—90
Spectrophotometric assays, 30, 85
Stability, lysate, 29
Staphylococcus aureus, 76
eye infections, 123, 125, 128, 129
meningitis, 76
Staphylococcus epidermidis
eye infections, 123, 125, 129
peritoneal dialysis, 136
Sterile medical devices, 165
Streptococcus, group B
eye infections, 123, 125, 128, 129
meningitis, 68, 70
Streptococcal septic arthritis, 74
Streptococcus agalactiae, 68
Streptococcus pneumoniae, 116
eye infections, 122, 123, 125, 129
meningitis, 68, 70, 76
Streptococcus viridans group, eye infections, 123, 125, 128, 129
Subarachnoid hemorrhage, 69
Superoxide anion, 9
Synovial fluid, 55, 112—114

T

Tachypleus tridentatus, 31, 59, 128
Tissue necrosis, 9
Tolerance, endotoxin, 9
Topical ocular medications, 131
Toxic effects of endotoxin, 8, see also Endotoxemia
Toxinometer ET-201, 32, 47
Trauma, abdominal, 137—138
Tumor necrosis factor, 9, 55
antibody to, 53
neurohumoral changes in, 54
Turbidimetric method
FDA regulation, 46
methods, 31
Typhoid fever, 8

U

Ureaplasma urealyticum, 98
Urethritis, gonococcal, 98—100
demography and clinical characteristics, 99
diagnosis in males, 100—105
epidemiology, 98—99
Gram stain in, 99—100
noncultural detection methods, 100
Urine assays, 10, 55, see also Gonorrhea
clinical usefulness, 87—92
false-negative, 88—89
false-positive, 89—90
screening methods, 91—92
literature review, 82—87
chromogenic substrates, 85—87
gelation endpoint methods, 83—85
spectrophotometric and turbidimetric endpoint methods, 85
Usefulness of assays, see also Sensitivity of assays; Specificity of assays
in meningitis, 74—77
in urethritis, 98—100
urine screening, 87—92
U.S. Food and Drug Administraion, see Regulation, federal

V

Vacuolization, 55
Vasoactive compounds, 10
Ventilation-perfusion abnormalities, 55
Veterinary applications, 156—157
Veterinary products, 165

W

Water contamination, 149—150
Water quality, hemodialysis, 144—145